Managing to achieve quality and reliability

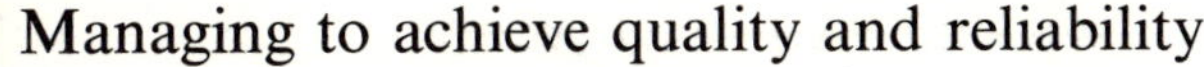

Managing to achieve quality and reliability

by Frank Nixon, C.B.E.

McGRAW-HILL

London · New York · St Louis · San Francisco · Toronto · Sydney · Johannesburg · Mexico · Panama · Singapore · Düsseldorf · Kuala Lumpur · Montreal · New Delhi · Rio de Janeiro

Published by
McGRAW-HILL Publishing Company Limited
Maidenhead · Berkshire · England

07 094223 4

PRINTED AND BOUND IN GREAT BRITAIN

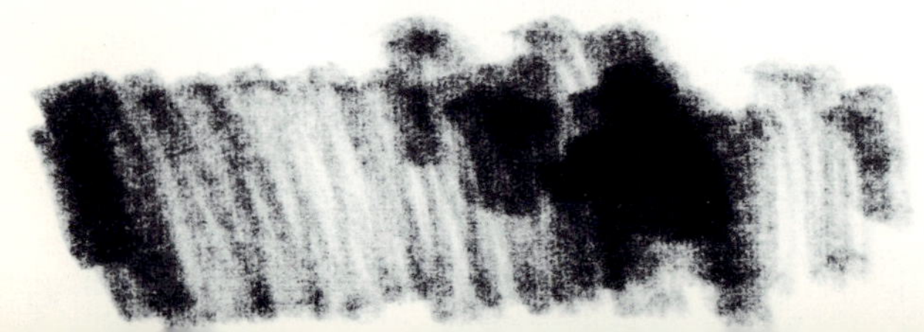

To Kerry

Preface

Behind every one of the tens of thousands of firms which comprise a nation's industry is an organization for the conduct of a manufactory or a service. Services themselves depend upon the products of many manufacturers. All of these manufacturing companies depend upon managers and engineers, this latter title being used in its original and truest sense of 'ingenious men'. It is upon the performance of these two groups of men that the continuing success of every industrial enterprise depends.

A study of the progress of many firms in many industries and in many countries has revealed a common pattern. Those which have lost markets by failing to maintain the competitive value of their products and services are those whose managements have not appreciated the importance of providing customer satisfaction. Others may have stayed in business, but have lost considerable sums of money through poor performance of their products. In both cases there has been poor rapport, the management failing to provide adequate facilities for their technical men. On the other hand, successful firms are seen to be those which never lose sight of their customers' needs, of the fact that the product or the service which they offer is their most direct means of communication with their customer, and that the integrity and the value of the product are matters of prime concern.

Yet firms in both categories may be using all the techniques of management which are available today.

It is the purpose of this book to show how the efficiency of every industrial enterprise, and its chances of survival, can be increased by paying due attention to the interrelated factors of customer satisfaction and satisfactory product. Consideration of these factors—of customer and product—leads to conclusions and hence to practices and concepts of great value.

Every organization is at once both supplier and customer. Major firms supplying the end-customer are themselves customer to hundreds or thousands of firms from which they obtain essential supplies and services. The success of the main firm clearly depends upon the ability of its management to involve all these firms, and all the people employed by them and by their own organization, in the end-purpose of the enterprise. Yet remarkably few companies recognize this, and take appropriate action.

Parenthetically, there are important lessons for managers in the Western World in the way in which Japanese companies enlist the support and cooperation of their employees. Admittedly, the traditional and inherited respect for authority is not so common in the West. Granted, the one union/one company practice makes things much easier in Japan. It is important to realize, however, that these are not the only factors. Japanese management offers in return security of employment, good planning and thorough training, and a company reputation for success in its particular field. This results in a conscientious effort on the part of every employee of every grade to help towards the success of the whole enterprise. For as long as Japanese managements are able to maintain this situation, they will possess an important competitive advantage.

The evolution of good and reliable products is the responsibility of technical and professional men, engineers and designers. These men cannot succeed, however, unless they are given adequate opportunity to apply their arts and mysteries so as to bring the end-product to the necessary level of satisfaction. Unfortunately, the current trend towards narrow specialization, and especially the 'professionalization' of management, its promotion to a pseudo-science which emphasizes everything except the importance of the customer and the product, have widened the gulf of misunderstanding between managers and engineers.

During the past few years thousands of British firms have benefited, to the extent of 4 per cent or more of their turnover, by adopting the principles of quality control. This has been due largely to the initiative of managers and engineers concerned mainly with manufacturing.

Few managements, however, are yet aware of the far greater potential value of the *reliability* of their products or services. Yet customer satisfaction depends, in most cases, far more upon reliability of performance than it does upon quality in the industrial sense. The costs of unreliability in our defence forces run into hundreds of millions of pounds annually. Nationalized organizations calculate their unreliability costs in millions. The return on the capital investment of countless smaller firms can be greatly diminished by the breakdown of plant. Yet examples

viii

are common of the failure of managements to provide their engineers with the chance to prove their work.

It is the prime purpose of this book to show how the efficiency of every industrial company, indeed its chances of survival, can be greatly improved by establishing a better mutual understanding of the roles and responsibilities of managers and engineers. An important matter is the essential planned system which must exist to coordinate the activities of all departments concerned with the conception and production of the end-product.

This 'system' receives little attention in books on management. Yet it comprises the footbridges between the activities which are shown in stark isolation on organization charts. This system provides a strong and flexible matrix in which can be embedded the industrial diamonds represented by the techniques of modern management, transforming them into tools of greatly enhanced effectiveness.

No excuse is proffered for the emphasis upon the practices of the aircraft industry. Despite the large size of modern concerns, they follow practices which were established when they were small companies, within the experience of many of today's executives. These practices were well-established before management was bombarded with advice, techniques, and nostrums for improved productivity. Their roots go back to the time when Britain was the 'workshop of the world', and the practices advocated in this book are applicable to firms of every size, in any industry.

FRANK NIXON

Contents

PART 4 CONVERTING CONCEPT INTO REALITY

Acknowledgements

The philosophy which runs through this book derives from experience with eminent engineers and managers who applied the principles of quality and reliability long before these words were in common use.

Grateful acknowledgement is made, in the first instance, to Sir Roy Fedden and to the late L. F. G. Butler, whose achievement in establishing the Engine Department of the Bristol Aeroplane Company as a world leader in its field deserves to be better known. There followed service under the late Lord Hives, who worked in a different way to build up the Rolls-Royce Company. These men transmitted knowledge and principles with a vigour and a discipline which it is hard to envisage in today's less rigorous climate.

More specifically in the fields of quality and reliability as they are now recognized, acknowledgement must be made to J. W. Lawrence, for his stimulating cooperation in developing reliable heat exchanger equipment for the RAF by cyclic proof testing. The late Victor Green, of the British Productivity Council, played the key role in establishing the British Q & R movement. Discussions with him and with H. W. Mander, of Automotive Products Ltd., a pioneer in the practical application of basic principles, helped greatly to mould the British approach.

Particular mention must be made of the generous help given by E. G. D. Paterson, a member of the original Quality Assurance team of the Bell Telephone System, and later its director. He traced the history of that pioneer team and showed how it was established, so that the pattern remains unchanged to this day. Discussion with John Riordan, who has done more than most to formulate the principles of quality assurance and to convey them to contractors to the US Department of Defense, has been equally valuable.

The cut and thrust of debate during the presentation of papers in many countries has been of great assistance. Those who participated are gratefully remembered. Particular mention must be made of Drs Edwards Deming and J. M. Juran, Leslie Ball, A. V. Feigenbaum, E. J. Lancaster, and Dorian Shainin, who have exchanged ideas ungrudgingly. Welcome confirmation of one's own viewpoint came during a memorable lecture tour of India, in the company of Professor K. Ishikawa, of Japan. The close identity of our ideas was remarkable.

Service on the Raby Committee, set up by the Ministry of Defence in 1968, offered a rare opportunity of seeing at first hand the management practices of large numbers of British firms, with products ranging from jerricans to nuclear submarines, by way of electronic devices, tanks, ships and aircraft. Many companies have been generous in their permission to draw upon their practices, and particular mention must be made of EMI (Electronics) Ltd, the Ford Motor Company Ltd, Marks and Spencer, and the Rover Company Ltd.

I am grateful to E. G. D. Paterson, Major-General B. G. Ralfs, and Air Commodore T. Wharton for their valuable comments and criticism of the draft treatment of the book, and to Jack Holmes for reading and commenting upon the manuscript. Sr. J. Garcia del Valle of Metalurgica de Santa Ana has been most generous in permitting reproduction of important parts of his company's quality policy manual. Rear Admiral W. A. Haynes, RN, gave welcome support to my views, by his pioneer work on quality assurance in the Navy Department of the Ministry of Defence.

Finally, if any of my colleagues are to be singled out for especial mention they must be W. G. Cook, whose rare ability to see the fundamentals of a problem was of enormous value when we worked together to establish the new approach, and Mrs M. Butler, who once again has worked without stint to produce an orderly manuscript out of chaos.

PART 1

The Basis of Successful Industry

1

Businesses that Succeed . . . or Fail

A business, if it is to continue to exist, must have a market. To have a market it must provide products or services which meet a need. Products usually involve the conversion of raw materials into useful form, by processes of design and manufacture.

Whether a business enterprise delivers a product or a service, it depends upon the work of numbers of people, whose effort must be marshalled and directed towards the ultimate objective. The organization of design and manufacture on the one hand, and the direction and motivation of personnel on the other, are two of the most important of the many essential activities of management.

For well over a century there has been a continuous and rapid increase in the size and complexity of industrial enterprises. This increase has been accompanied by a growing literature on the subject of management. This literature has passed through several phases, from John Stuart Mill's broad economic survey of industry, the practical 'workshop boss' approach of F. W. Taylor, Fayol's administrative concept, on to the present day emphasis upon human relations, organization, delegation or the alternative of centralization, with advanced techniques of operational research and quantitative techniques of decision making.

In this technological age, the manager is required to have at least a working knowledge of a large number of techniques applicable to a number of different areas of activity. There is no lack of information and assistance available to him on the specialities which have been developed. What is surprising, however, is the general failure to recognize, or at least to emphasize clearly, the importance of the customer, the product, and the system of communications without which no organization can be effective. With the exception of some of the earlier writers of this century, notably Emerson, the importance of these is usually found hidden in the general term 'management plan'.

Yet without clearly defined objectives, and a thorough and detailed system of communications, comprising clear and accurate instructions of what must be done, with records to confirm that it has been done, no industrial organization, no matter how sophisticated and up-to-date its managers, can hope to achieve efficiency and success. A surprisingly large number of firms, of all sizes, have been found to have no plan along these lines.

In the more successful companies, these factors have long been recognized and implemented. The current emphasis upon quality and reliability has focused attention upon the importance of these factors. This in turn has enabled many more firms to appreciate and to adopt that planned system which has been so long established in the older firms that it has come to be taken for granted.

The Purpose of Business

The Harvard School of Business Administration had been in existence for 50 years before it was realized that it had yet no definition of the business whose administration was the subject of study. The occasion was the 50th anniversary convention. From the ensuent cerebration there was evolved the classic definition: 'Businesses exist to create and deliver value satisfactions at a profit.'

Managers not having the advantages of the formal education provided by one of the world's leading institutions may be forgiven for taking longer to realize the *raison* for their *d'être*. They have not been helped by many people who, having acquired an understanding of some one or more facets of the factors influencing productive output, have imagined that this was the whole of the complex art of management. Some of these people have offered advice and instruction to managers whose practical experience far exceeds theirs. Politicians and journalists add their quota of criticism and exhortation. Today's manager is also the recipient of a spate of leaflets advertising courses and seminars dealing with some 'newly-discovered technique' of management.

As a result, managers tend to become imbued with a sense of guilt, a feeling that they are falling behind the times. Looking inwards towards the organization which is allegedly in need of care and attention, they are diverted from external factors which may be of greater importance. With so many remedies proffered to them, they fail to make a proper diagnosis of the overall situation. They adopt the nostrum which is currently most popular, feel that by doing so they have discharged their responsibilities, and relax. Far too often the remedy they adopt is not

4

appropriate to their problem. The hoped-for improvement does not materialize, disillusionment follows, and the adoption of other management methods which would have been of great value is retarded.

Managers who practised their art in the early days of this century, before management had been converted into a 'science', were more fortunate. Although those aids to management which are demonstrably of value were not available to them, they had a better understanding of the overall aims of their enterprise, and of what they themselves had to do in order to succeed in business, than is possessed by many of today's 'professional' managers. Some of them have indeed bequeathed to those who followed after them a philosophy and practice which is the basis of their firm's continuing success.

Undoubtedly things were simpler and issues could be seen more clearly when companies were smaller than many are today. It was easier, therefore (and still is in many successful smaller firms), to maintain in perspective the responsibilities which management must meet if their organization is to succeed.

Companies which survive the ups and downs of political and economic changes can be recognized as those which consistently satisfy their customers. Those companies which are more successful than their competitors are the more efficient. It is likely that both they and their competitors will be benefiting equally from the application of good methods of management. The leaders will be those who realize that the product or service they offer is the means by which customer satisfaction is achieved. They see clearly that the product of their efforts is the tangible means of communication with their market.

Management and Engineering

The practical and essential business of evolving satisfactory product is of direct and vital concern to management. Yet it is largely ignored in books on the subject. Evolution of product, its design, and its manufacture, are technical matters, requiring the application of engineering talent. Unfortunately, engineering is neither understood nor appreciated by many managers, just as management is ignored as an essential subject of study by many engineers.

In consequence, we find too many managements failing to realize the importance of the technical departments of their organizations. They find it difficult to understand why they should allocate what seem to them to be excessively large sums of money for design and development activities.

The prevalence of the attitude of mind which fails to appreciate the importance of the engineering function has been shown by the poor response to the Feilden Report on Design. More concrete evidence is provided all too frequently by the examples which we have seen during the past few years of breakdowns of new technical projects. Costs have been piled upon costs, due to lost revenue, lost reputation, and the work necessary to put things right. These costs have been due, in almost every case, to failure to appreciate the importance of reliable product to the 'value satisfaction' of the customer, and of the procedures which must be followed to achieve the necessary reliability. In many cases, the engineering resources which are essential for the needs of the project are not made available. In other cases existing resources are not used to the full, in misguided efforts to save money.

> Fortunately, there are brilliant exceptions. The success of the Vosper Thornycroft Group has been described as due to its management policy 'of high productivity and the successful integration of a highly specialized technical team with an equally specialized sales force'.
> In addition, the Group's Managing Director, John Rix, says, 'In a technically-based industry you must head your sales force with trained engineers.'[1]

Engineers for their part must accept some part of the responsibility for the prevalent state of affairs. They have failed to explain, in terms which will be understandable and convincing to management, what must be done to ensure that reliable product will be achieved.

Unfortunately, in these days of excessive specialization, there is little appreciation by managers of the responsibilities and the function of engineering, or by engineers of the worries which beset management. An effort must be made to bridge this gulf between two of the essential parties to every industrial enterprise.

For the present, it will be sufficient to use the term *management* to describe the total activity in the enterprise which comprises the work and responsibilities of all its managers. The *managers* are those whose job it is to help others to understand their responsibilities, and to be able to discharge them. Management, therefore, encompasses people ranging from the foremen who manage their teams of operatives, to the most senior executive in the company. Whether *management* is used to describe the total activity, or that of the upper echelon that makes decisions affecting the future of the company, will usually be clear from the context.

Engineers are taken to be all those who are concerned with the evolution of the product, or the implementation of the service. In a particular organization they may be some or all of those concerned with the market research, project, and product design, confirmation of the design, planning, and execution of the manufacturing process, selling, and servicing the product.

Nor must we confine our interests to manufacturing industry. Those industries providing services—railways, bus companies, transport fleets, public utilities, the defence forces, distributors, and retailers—all have the same obligation to provide value satisfaction. (This is especially topical at a time when we expect adequate defence for a continually diminishing Vote.) In all of these organizations there is management, and there is engineering. It is perhaps not generally known that the late James Wright, who did so much to implement the quality/value policy of the late Lord Marks of Marks and Spencer, was trained as an aircraft engine engineer.

It will be shown later how the thinking of both managers and engineers is greatly clarified by concentrating on the aim of providing customer satisfaction. By tackling first things first—the creation and delivery of products of adequate quality and reliability at a competitive price—it becomes easy to identify the major areas of weakness in the organization. In many cases, these weaknesses can be eliminated with little effort beyond the changing of attitudes of mind. This in turn will enable more effective selection to be made of the management aids which could contribute to the efficiency of the organization. Thus will executives be armed against the zealots who would have them accept the latest technique simply because it is new, or dressed up in a new guise. Whether or not it is likely to help towards the achievement of the main purpose is seldom considered.

Customer-Conscious Management

Companies which survive slump and boom, buyers' and sellers' markets, are those which provide better value to their potential customers than can be obtained from their competitors. Naturally, when demand exceeds supply, inferior products will find a ready market, but here we are concerned with the longer-term view.

The financial pages of most newspapers give the impression that the main preoccupation of business is to make a profit. The means by which this is done—the creation and delivery of value satisfaction by the timely appearance of appropriate product at an attractive price—is barely noticed, save when some new enterprise, by brilliant engineering, achieves a newsworthy breakthrough.

Emphasis upon profit, and the ensuent regard for productivity of manufacture, divert attention from the essential aim. This should always be to satisfy the customer. In a sellers' market it is difficult to resist the temptation to go for maximum profit, to make a 'quick killing'. Yet this is the very time that there should be greatest effort to satisfy the customer. 'Work hardest to please your customer when he matters least' is excellent advice proffered by Shri Arunachalam Jr, Vice-President of Tube Investments of India.

John Ruskin said that people who consider only price, and not value, are the lawful prey of those who purvey shoddy goods. Despite the human tendency to succumb to the temptation of a bargain, however, buyers are rapidly becoming more selective and better informed. The market possesses a mass intelligence which will, sooner or later, enable the customer to know when he is being cheated. Housewives are notoriously able judges of value, and they will shop around the supermarkets to seek out the 'loss leaders', the well known proprietary goods which are being offered temporarily at a discount by different stores.

Consumer protection organizations publish tables of comparative values of packets of detergent. They educate their readers towards a more critical appraisal of the quality and reliability of functioning products. It is possible, too, that advertising on television may be having a similar effect. With so many manufacturers claiming that their product is the best, the potential buyer soon comes to the conclusion that there can be little difference between them, and he makes his choice on the basis of 'most for the money'.

On the other hand, with functioning products, motor cars and kitchen equipment for example, a reputation for reliability and good back-up service is increasingly the factor determining the ultimate choice.

An extreme case of value dissatisfaction due to shoddy goods put on the market at inflated prices by unscrupulous vendors has been adduced as one reason for the serious riots in Washington, DC, in 1968.

Fortunately, not all dissatisfied customers are driven to such violent reactions. In the long run, however, products which do not satisfy cease to have a market. The company which does not strive continuously to provide value satisfaction will, sooner or later, find itself out of business.

The great majority of enterprises, however, have the sincere intention of staying in business. Nevertheless, a good deal of cynical misunderstanding exists about business enterprise, not least among people concerned with the financial and marketing aspects of commerce.

At gatherings of bankers, merchants, and exporters there have been criticisms of manufacturing industry. It has been alleged that manufacturers tend to follow a practice of 'planned obsolescence'. Perhaps because of Vance Packard's cynical exposé of opportunist salesmanship, it is of little avail to try to explain that it would be an extremely difficult technical task to design most types of product down to a minimum life. (See chapter 9.) The only really successful instance of planned obsolescence is women's hats, with their clothes coming a close second. Here, the tradition of seasonal change is the factor. Fashion, not technical improvement, is the criterion.

Careful examination of many instances of alleged 'planned obsolescence' shows that it was not deliberate intent, but neglect of the principles which will be described later which was the main factor.

There is, of course, some justification for the allegations. On the whole, instances of deliberate lowering of quality, durability, or reliability are in the minority, and any success they achieve for their perpetrators is usually short lived. For example, it has been stated publicly that some articles of electrical domestic equipment have been 'rationalized' to lower the cost of manufacture. This rationalization has resulted in there being fewer sub-assemblies, in a toaster for example, which are now welded together where previously they were bolted. In consequence, it is no longer possible to replace a small unit component such as a broken spring. If such a component fails, a much more expensive sub-assembly must be bought. Initial cost may have been reduced, but 'total life cost' will have been increased.

It would probably be wrong to accuse the manufacturers of a deliberate intention to raise the total cost to the customer. It is more likely that they had not appreciated the fundamental importance of providing maximum value to the customer.

Good examples of attempts to offer better 'total life' value have been provided by some motor car manufacturers and most aircraft engine builders. These have realized that maximum life, minimum cost of replacements, and minimum costs of maintenance are important factors in the total cost to the user throughout the useful life of their products. They know that their informed and educated customers are coming to realize this, more and more. Hence, in the automobile field, we see a trend towards longer guaranteed lives (note Chrysler's 5-year or 50,000 miles warranty). Easily replaceable wings, economically priced engines, brake assemblies, gearboxes, shock absorbers, with standardized fitting charges, are offered by some motor car manufacturers, notably the Ford Motor Company and the Rover Company in Britain. It can only be a matter of time before the others follow suit. Aircraft engine manufacturers offer guarantees of maximum spare parts costs to their airline customers, and work constantly to extend the lives of major components of their engines, in order to increase their value to the customers.

Examples such as these show the rights which a customer may expect of his supplier. Sooner or later they must lead to the emergence of a demand for better value and to an appreciation of total life cost in all other markets.

Product-Conscious Management

Concentration upon the aim of satisfying the customer leads to the realization that it is the end-product of the enterprise, be it hardware or a service, which is the vital factor in the implementation of the policy. The impact is even greater when it is seen that the end-product is the means of communication between the supplier and his customer.

The success which has been achieved by Japanese makers of cameras and builders of ships may be quoted as examples. These manufacturers have made considerable inroads into existing markets, by the attractiveness of their products, by offering better value.

In the case of cameras, this has been accomplished by designing-in the features which it was felt would appeal to the customer—better view-finding by through-the-lens reflex; better and more convenient exposure metering, also through-the-lens; automatic exposure control; inter-changeable ranges of lenses—all at highly competitive prices.

Bigger and bigger ships have been built, quickly, cheaply, and on time, by introducing into shipyards the techniques of aircraft building. The aim, all the time, has been to provide the customer with better value than he can obtain elsewhere, with a product offering more economical operation and greater profits.

This aim requires that the right product for the market shall be identified, by good market research. It must be designed to meet the requirements of the intended user, and the designer must be helped to study and to understand the conditions of use and environment. The design must be thoroughly proved by adequate development testing. It is up to senior management to allocate the funds necessary to cover these functions. It is also a responsibility of senior management to ensure that the available forces, of design and development engineering, are adequate for the purpose.

> A telling example of false economy is provided by the turbines of the *Queen Elizabeth II*. Lack of testing, and of testing facilities to establish the soundness of the design, meant that the presence of resonant frequencies in some stages of turbine blades remained undetected. Failures occurred early in the life of the ship. They

attracted a great deal of unwelcome publicity, and large financial losses were incurred. It is highly probable that these could have been avoided had a fraction of the total amount lost been spent on what should have been looked upon as essential development testing, before the turbines were installed in the ship.

The proven design must be manufactured efficiently and correctly. There must be good after-sales service to minimize the effect upon the customer of any troubles which might arise in use, and to reap the benefit of service experience.

These activities amount, in the aggregate, to the present-day concept of quality and reliability. They were well understood by managers who were practising long before the current titles were conceived.

In later chapters, there will be described an account of basic principles and practices of product development which were established before the advent of modern management techniques. By combining them with these techniques, management can avail itself of an overall approach by means of which it can be assured that it will have done its best to satisfy its customers and remain ahead of the competition.

References

1. *The Engineer*, 27 November 1969.
Passim Urwick, L. *The Golden Book of Management*. Newman Neame Ltd., London 1956.
George, Claude S. Jr. *The History of Management Thought*. Prentice-Hall, Inc., N.J. 1968.

2

Why Quality and Reliability?

Four hundred years ago Richard Hakluyt showed an appreciation of the importance of good product which was perhaps more clearly understood than it is today. He proffered the following advice to the exporters of his time:

> To take with you those things that be in perfection of goodness. For as the goodness now at the first may make your commodities in credit in time to come! so false and sophisticate commodities shall drawe you and all your commodities into contempt and ill opinion.

In the 16th century, commercial enterprises were small. The men in charge led teams of individuals who realized that they were in business to satisfy the needs of their customers. Moreover, the organization usually existed to supply a local need, and the customer was often able to visit the manufactory in person. An intimate relationship existed between supplier and customer. Each individual craftsman realized that the quality of his workmanship was under the direct and immediate scrutiny of the buyer.

Today, with the greatly increased size and complexity of industrial firms often having worldwide interests, with the work of many individuals reduced to simple repetitive operations, and with a proliferation of management 'techniques', it becomes more difficult for workers and management alike to keep in mind the essential aim of the organization which provides their employment. This must be, at all times, to provide products which will excel in the competition with other similar enterprises by better satisfying the available market. In the long run, this is

12

the only way in which a firm can stay in business and continue to provide work for its employees.

For their part, aided by the advice offered by the publicity agents of competing enterprises and by organizations which have come into being for the protection of the individual customer, potential buyers are much better informed than they were only a few years ago. They are becoming more critical and selective, and as a result the quality and reliability of products and services, both of which are major factors in the buyer's assessment of value, are once more assuming the importance which was theirs in less confused times. Concentration upon the quality and reliability of products provides a short cut back to the primary elements of management upon which most of today's successful undertakings were founded. It is doubly fortunate that at the same time Q & R offers opportunities for making considerable improvements in productivity, efficiency, and competitiveness.

> A few years ago the author was visiting the quality manager of a company producing capital goods on a large scale, to be shown how it was hoped to introduce quality control. He was called into the office of the production director, who reproached him for appearing to work against the drive to increase exports because 'We just can't afford "quality"—it will increase our costs'. The director instanced a recent meeting which he had held with shop stewards. He had been insistent that the company required better quality of work from the men. 'All right then', replied the shop stewards, 'if you want better quality from us, we want better money from you.' And there the matter rested for the time being.
>
> Asked what was the 'better quality' that he was demanding the production director was at a loss to reply. It was suggested that as the workers had already shown themselves capable of producing the majority of parts to an acceptable standard, he would have been better advised to ask for more parts to be made like the good ones.

This story typifies the misunderstanding which is still prevalent in large sectors of the industry of many countries. It is useful as helping us to identify some of the more important weaknesses which are commonly found to exist in the practical application of the principles of management.

We see, for example, the busy manager's natural suspicion of something which has become the subject of zealous campaigning and, what is worse, which seems to demand expert knowledge of a new technology. There is the all too common interpretation of 'quality' solely in its sense of 'degree of excellence'; there is the implied assumption that it is the production operatives who are responsible for all failures to achieve the requisite quality; above all, there is a total failure of communications through the omission of any definition of terms.

'Q & R' is Basic

A notable feature of the quality and reliability approach, or 'Q & R' as it has come to be called in Britain, is the way in which study of its implications concentrates attention upon the product and upon the customer. By focusing the interest of all who are engaged in industry upon the achievement of satisfactory product and satisfied customer— ends with which no one can quarrel—a spirit of involvement can be aroused resulting in substantial improvements in productive efficiency and, above all, in increases in competitiveness of product through enhanced value. An important further advantage is that the improved perspective which results from consideration of these aims enables us to put in their proper place the increasing number of techniques and methodologies which have tended to obscure rather than clarify the main issue. The Q & R movement has suffered much from confusion between means and ends. Fortunately, in its current British form, the end objective is so clear that it is not difficult to prevent the means from assuming undue importance.

Ends before Means

The kernel of Q & R is contained in the definition which the Harvard School of Business Administration evolved at its 50th anniversary meeting and which has already been quoted.

For the moment let us examine the *end*—'value satisfactions at a profit'. The *means* of creating and delivering these value satisfactions, which are the fundamental aims of the quality and reliability movement, will be discussed more fully in later chapters.

The customer, without whom no industrial enterprise can survive, judges his satisfaction with a product by the extent to which it meets his needs. It is all too often forgotten that each customer is an individual, with his own unique set of tastes and wants. It is therefore he, and only he, who can be the final judge of the measure of his own satisfaction, and he makes his assessment in terms of value: is it the best that he can get for his money? This in turn involves competition. The housewife seeks to discover which market offers the tin of peaches of her choice at the lowest price. The family man buying a motor car examines the competing models which are within the scope of his pocket, applying a

14

whole range of criteria which will vary in individual importance according to his personal requirements and desires. These will include size, depending primarily upon the size of his family, then upon the size of his garage space, and finally his own whim; performance; economy; and an indefinable aesthetic appeal. The more discerning potential buyer will consider the reliability and service support offered by different manufacturers, and here he will be influenced by his previous experience or by the reputation already established by the manufacturer.

This explanation provides an answer to those who say that a Rolls-Royce is the highest quality of car available. To the man with no more than £700 to spend, a popular 900 cc car may be for him the better, indeed the only, choice.

At the same time, we are provided with the answer to those people in industry who become puzzled by the acceptable quality level of the statistically minded quality control professional. They ask, 'What is the quality level which we must achieve?'

The short answer, as applied to the end-product, is 'something discernibly better than that of your nearest competitor'. Whether we like it or not, we live in a highly competitive world and it is we, the individual customers, all of us, who impose this competitive criterion upon our suppliers. Since it is we who are in turn involved as suppliers to yet another set of customers, who are we to complain?

The increasingly critical and selective market for consumer goods, which is a direct consequence of a high standard of living and of the existence of many competing products, is one of the biggest factors compelling management to take a serious interest in quality and reliability.

Management may with some justification feel hurt by what they feel to be unfair criticisms of the quality and reliability of their goods. The motor car, for instance, is a much more complex machine than it was, say, 30 years ago. Yet it is faster, with much better acceleration, better economy, greater comfort, and greater dependability. Today's motorist is apt to forget, if his experience goes back far enough, what were common troubles only a few years ago. These included difficult starting, flat batteries, flat spots in acceleration, poor acceleration, poor road holding, frequent decarbonizations and rebores, noise, ineffective windscreen wipers, no windscreen washers. It is now taken for granted that these troubles should not exist, so the customer turns his attention to minor rattles, draughts, misting of windows, faulty door catches, paintwork, and trim.

If he experiences any of the former troubles, as regrettably he sometimes does, the dissatisfied customer is much more vocal than he was formerly. He has become a force to be reckoned with, and the manufacturer of the car concerned will not achieve his hoped for sales.

After a detailed investigation, a large American corporation came to the conclusion that about forty potential customers became aware of each complaint by a single customer. Hence, each complaint had a far greater adverse effect upon potential sales than might be assumed from the individual cases reported.

And so, whether industrialists like it or not, customer satisfaction must be the first and most important ingredient in any plan for success.

Definitions

As we are here concerned with industrial organization, it will be appropriate at this stage to become a little more specific about the terms employed.

The European Organization for Quality Control has evolved the following definitions, which have been widely accepted:

QUALITY The quality of a product is the degree to which it meets the requirements of the customer. With manufactured products, quality is a combination of quality of design (*q.v.*) and quality of manufacture (*q.v.*).

QUALITY OF DESIGN The value inherent in the design; a measure of the excellence of the design in relation to the customer's requirements.

QUALITY OF MANUFACTURE (quality of conformance) A measure of the fidelity with which the product, taken at the point of acceptance, conforms to the design.

RELIABILITY The measure of the ability of a product to function successfully, when required, for the period required, in the specified environment. It is expressed as a probability.

These definitions were derived by and for professionals, so a word of explanation may be necessary for those not so directly involved. Historically, the inclusion of 'quality of design' as a definition represented a great advance in the thinking of those who had regarded control of quality as an end in itself. Consideration of the dilemma facing the man who is about to buy a motor car will show that *quality of design* applies to this and indeed to every product, in whatever price bracket. So too will *quality of manufacture*. This same man can carry out his own visual acceptance inspection of his motor car before he takes delivery. His examination, however, can do no more than satisfy him that in outward respects—of furnishing, trim, paintwork, and functioning of primary controls—the car is satisfactory. It can tell him nothing about the reliability and dependability of performance which he will

16

get from the car of his choice. Only future experience can tell him that.

This is why the definition of reliability contains what many may find to be a puzzling clause—'expressed as a probability'. A great deal of confusion has arisen about this, and it will be examined more fully later. It is sufficient at this stage to say that 'probable reliability' is hardly a saleable commodity. A buyer demands something more convincing, and in the case of an individual purchaser of a motor car this will usually be the reputation already established by the company producing the car. If reliable service is backed up by a meaningful warranty, so much the better.

All who are involved in industrial enterprise of any kind must have as their main objective the task of satisfying the customer. If they do not, then eventually someone else, by producing a product more attractive, or as good but cheaper, will win the orders. The firms which stay in business are those which realize this fact of commercial existence. In times of excess of demand over supply, it is all too easily forgotten. But customers have long memories and, with the certitude of an immutable law of nature, nemesis will eventually overtake those managements which ignore this fact.

The importance of quality of products, of management, of workmanship, was given supreme emphasis by Vice-Admiral H. G. Rickover, USN, who showed how vital it was to achieve far higher standards of quality if nuclear submarines were to be produced to perform safely and satisfactorily. The success achieved in this field was so impressive that the British Ministry of Defence (Navy Department) was quick to realize the potential value of the quality and reliability approach to the entire shipbuilding industry. Today, shipbuilders are among the first to acknowledge the important beneficial results which have accrued from the adoption of methods of quality assurance.

The customer is the man who must receive satisfaction from those supplying him with goods or services, if he is to remain their customer. In this context, it must be realized that the customer may be the next man in a production line who receives part-machined pieces from the man before him; or the next department receiving parts for further processing; or the employer who buys the services of his employees; or the prime contractor buying materials or sub-components from any of his suppliers (and often there may be between 500 and 1000 of these); or the ultimate buyer of the end-product. Hence, 'The customer is boss' is a healthy concept which has been adopted as a slogan by some successful smaller-sized American companies. This slogan could well take precedence over all the others which are frequently seen on notice boards in workshops and offices.

Value

We have seen, indeed we have all as customers experienced at first hand, how assessment of value is based upon the two factors of quality and reliability as already defined, for the price demanded. Hence, value can be defined as

Quality and reliability
<hr>
Price

It is clear that by increasing the quality and/or reliability of a product, and/or by reducing its price, its attractiveness, i.e. its value to the customer, can be increased. What is not so obvious is that increases in product quality and reliability have, more often than not, a double advantage which is as yet only dimly appreciated by all but a minority of managements. Improving quality and reliability can do much more to increase the value of the product than can be effected solely by reducing its price.

Quality, in the sense of quality of design, costs little to achieve, yet it can enhance greatly the appeal of a product to the customer. Quality of conformance with the design has been shown beyond all possible doubt to be a sure measure, not only of increased customer satisfaction, but of increased efficiency of manufacture as well. Hence, quality improvement usually reduces costs. Improved reliability often leads to increased sales, which again enables prime costs to be reduced, while at the same time it reduces 'total life cost'.

In-Plant Quality Costs The costs of ensuring that only those products which have been manufactured to specification reach the customer have been called 'quality costs'. They have also been referred to as 'gold in the mine' by Dr J. M. Juran, because they represent a manufacturing cost which is capable of substantial improvement.

Quality costs comprise the losses due to:

(a) Unsatisfactory parts which must be scrapped as unacceptable.
(b) The reworking or correction of those unsatisfactory parts which can be made acceptable.
(c) Inspection to ensure that unsatisfactory parts do not reach the customer.
(d) Such preventive action as may be being carried out to avoid the production of unsatisfactory parts.

The magnitude of these costs is seldom appreciated by management. The main reason for this is that they are usually hidden in the overheads, because the cost accountants have not been asked to extract them. Typical figures for an 'average' firm of 1000 employees are given in Fig.

18

2.1. They can safely be used as a basis for the first approximation to the quality costs of a firm of any size, by scaling them pro rata to the number of employees.

<table>
<tr><td colspan="2">1. BEFORE ADOPTING QUALITY CONTROL</td></tr>
<tr><td>Total number of employees</td><td>1000</td></tr>
<tr><td>Total turnover</td><td>£4,000,000</td></tr>
<tr><td>Number of direct operatives</td><td>400</td></tr>
<tr><td>(i.e. directly concerned with producing, excluding supervision, inspection, shop labourers, services)</td><td></td></tr>
<tr><td>Number of inspectors</td><td>50</td></tr>
<tr><td>Laboratory or material test</td><td>7</td></tr>
<tr><td>Product test Omitted, as this will vary widely according to the type of product. In many cases there will be no testing of the final product.</td><td></td></tr>
<tr><td>Costs:</td><td></td></tr>
<tr><td>Inspection, at £1200 p.a. average, plus 100% overheads</td><td>£120,000</td></tr>
<tr><td>Laboratory or material test, at £1800 p.a. average, plus 100% overheads</td><td>£25,200</td></tr>
<tr><td>Scrap and rework, assumed to be 7% of turnover</td><td>£280,000</td></tr>
<tr><td>Total quality cost</td><td>£425,200</td></tr>
<tr><td>Quality cost as percentage total turnover</td><td>10·6%</td></tr>
<tr><td>Quality cost per direct operative, p.a.</td><td>£1062</td></tr>
<tr><td>Scrap and rework cost for direct operative</td><td>£700</td></tr>
<tr><td colspan="2">2. AFTER INTRODUCING QUALITY CONTROL</td></tr>
<tr><td>Costs:</td><td></td></tr>
<tr><td>Quality Control Department Comprising QC manager and staff of 6, at £2000 p.a. average plus 100% overheads</td><td>£28,000</td></tr>
<tr><td>Inspection, reduced by 25%, say</td><td>£90,000</td></tr>
<tr><td>Scrap and rework, at £250 per operator per annum</td><td>£100,000</td></tr>
<tr><td></td><td>£218,000</td></tr>
<tr><td>Saving per annum, or 5·2% of turnover</td><td>£207,200</td></tr>
</table>

Fig. 2.1. Quality Costs and Savings, for a Typical Manufacturing Company.

These quality costs amount, in the average case, to not less than 12 per cent of total turnover, a figure comparable to the labour costs of many firms (Fig. 2.2). This figure by no means represents the total cost of inadequate quality and reliability, as it includes nothing for warranty claims, service engineering, customer placation, or loss of reputation due to customer dissatisfaction. Nor is it possible to include the cost of that effort to achieve the requisite quality which is, or should be, undertaken by every department of the organization, including design,

19

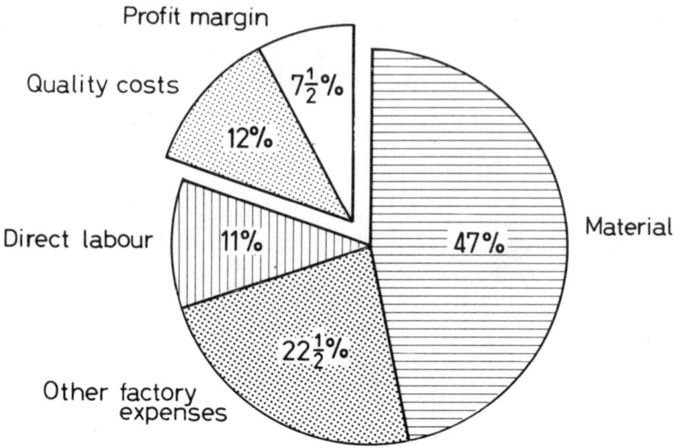

Fig. 2.2. The Constituents of Selling Price.

development, production engineering, operation planning, buying, manufacturing, sales and service, for example. For reporting to top management, and to focus attention upon the immediate and easily achievable target, they are taken as including only the costs of the more obvious operations, as indicated in Fig. 2.3.

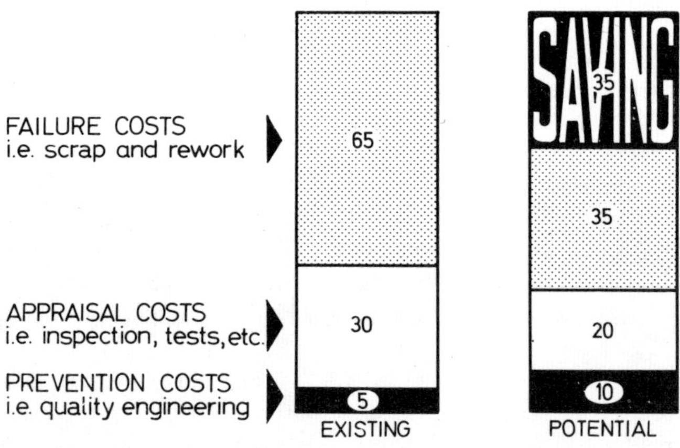

Fig. 2.3. Quality Costs: the Gold in the Mine.

This diagram shows the typical build-up of quality costs. It will be seen that the cost of scrap and rework is about twice as important an item as the cost of inspection. Yet examples are numerous of cases where

20

recognizably poor quality performance has been tackled by increasing the inspection effort!

The case for the adoption of quality control depends upon the argument that it is always more efficient and economical to get the job right first time—to take such action as will reduce the risk of error at the point of production—rather than to depend upon inspection to sort good parts from bad, after the mistake has been made. So many examples have been quoted of the savings achieved by companies of all sizes and kinds, in many different countries, that it may be taken as firmly established that quality costs are capable of appreciable reduction. Such savings can be applied to a lowering of prices, or to enhance an inadequate profit, or to increase productivity, or to some combination of any or all of these. Marks and Spencer Ltd, who have during the past few years provided a classic example of the effective application of quality control, quote cases in which their suppliers have increased their output by as much as 30 per cent, by eliminating defective work.

On the other hand, reductions in price deriving solely from improved methods of manufacture, without the benefits of reductions in quality costs, can seldom be of the same order as the increases in value following improvements in quality and reliability.

It follows therefore that, in general, value can be increased more, and much more easily, by raising quality and reliability than by reducing price by any amount which might conceivably be achievable.

During Quality and Reliability Year, 1966–1967, about 8000 firms took part. Many of them reported reductions in manufacturing cost amounting to 3–5 per cent of selling price, by eliminating causes of scrap and rework and by reducing the amount of direct inspection by introducing methods of in-process quality control. In aggregate, the national savings were estimated at £50 million. But, if the whole of manufacturing industry adopted the basic principles of Q & R, a saving of ten times as much, or £500 million, is felt to be a not unrealistic possibility.

Savings ensuent upon improved quality of conformance can be computed fairly easily and quickly, in terms of reductions in the cost of ineffective work on spoilt parts, and in the cost of unnecessary or ineffective inspection. The value of improved reliability is not so easy to determine.

The Cost of Unreliability In Britain, the pioneer effort to estimate the costs of unreliability was made by the Royal Air Force. It has been estimated that the failure of equipment to live up to its desired performance costs the RAF about 40 per cent of the total annual Air Vote of about £500 million.[1] This is taken to be the cost of the spare parts which must be held in stations all over the world, as an insurance against emergency needs; the cost of the staff necessary to carry out routine replacements and voluntary and involuntary maintenance and repair; the cost of scheduled overhauls necessary at uneconomically short intervals of time; the cost of modifications in design to overcome weaknesses apparent in the design as commissioned, and so on. Being a military service, however, nothing can be included, in financial terms, for the cost of an abortive mission, a lost battle, a lost strategic exercise, or even a lost war.

Following the example of the Royal Air Force, the National Coal Board estimated the importance of the reliability of its coal-winning equipment. In this case, it could include the costs of coal not won by equipment which was not functioning. It was found that 40 minutes per shift of 8 hours were lost through unreliability of mechanical/electrical equipment. A reduction in down-time of one and a half minutes per shift would result in an increase of income of £1 million per annum.

Generally, a great part of the consequential costs of unreliability cannot be computed—indeed they appear in no balance sheet anywhere.

> For example, consider the case of a company which has won important export orders in the face of intense competition. The customer has imposed stringent penalties for late delivery. There has been the usual rush to complete the job on time, and the equipment has been crated and loaded and the lorry sent on its way to the docks with comfortable time in hand. En route, the engine of the lorry suffers a breakdown through a burnt-out exhaust valve, and the shipment is missed. Penalties are incurred and a marginally successful company receives a black mark.
>
> Who is to blame? The contractor, for faulty maintenance of his fleet of trucks, or the lorry supplier? In any case, the only recompense the contractor *might* receive, after correspondence costing far more than its worth, could be a free replacement exhaust valve. Yet the losses incurred by him could be incalculable. In addition to the penalty incurred, he might well lose the repeat order.

The discriminating customer, more especially the one concerned with large purchases of capital equipment, e.g. a fleet of aircraft, or lorries, or a power station, is becoming increasingly concerned with the total cost throughout the useful life of the equipment he purchases. In the case of complex military equipment, this total lifetime cost might well be several times the prime cost. It is probable that, in the appreciation of the importance of total cost, the lead was taken by aircraft engine companies, who guarantee to their commercial airline customers that the cost of using their engines will not exceed a certain figure per hour. The importance of reliability as contributing to this total cost was recognized by Premier Krushchev who, in 1964, drew the attention of the USSR State Committee on Aircraft Engineering to the long lives between overhauls being achieved by British aero engine companies. Russian engineers were required to improve their performance. This concept has been pursued by the American Department of Defense, which is beginning to take a critical look at the total costs of possessing any particular item of equipment ('life cycle costing'). The Ministry of Defence (Navy Department) in Britain is designing incentive contracts which will take into account the value to the customer of improved and extended reliability, and which will reward the contractor according to his performance in this respect.[2]

An early instance of recognition of the importance of reliability to the customer was provided by the Chrysler Corporation, which was the first to offer a warranty (albeit limited at first to engine, gearbox, and transmission) to those who bought its cars and who maintained them properly, of 5 years or 50,000 miles, whichever was accomplished the sooner.

It is gratifying to note that this trend has led to the doubling of the traditional guarantee of one year, by more than one British and German car manufacturer.

Instances of the value of an interest in quality and reliability could be multiplied. Since we are concerned here with people in industry, who may be considered to be already well-informed, it will be sufficient to offer in summary form the additional reasons why they should be taking an interest in quality and reliability.

The Advantages of Q & R
In 1960, at the Fourth Annual Conference of the European Organisation for Quality Control, which was held at Church House, London, the author adduced the following advantages:[3]

(a) Reduction in cost of scrap.
(b) Reduction in cost of rework of defective parts.
(c) Saving in material, which may be in short supply.
(d) Saving in productive labour, better utilization of existing skilled labour.
(e) Saving in inspection costs.
(f) Saving in cost of dealing with customer complaints.

(g) Reduction of losses due to discount on 'seconds'.
(h) Lowering of charges to guarantee and warranty fund.
(i) Fewer delays and stoppages.
(j) Shortening of production cycle time, resulting in earlier deliveries, and reduction of capital locked up in work in progress.
(k) Easement of difficulties and problems of production control caused by defective work.
(l) Increase in customer goodwill, due to greater reliability and enhanced value of product, leading to increased business and, in the ultimate, continued survival in the face of competition.
(m) Reduction of internal friction within the organization.
(n) Improvement in morale of operatives due to consciousness of a job well done, and closer identification with the end-product, leading to increased pride in their workmanship.

It is more than adequate supporting evidence of the soundness of the Q & R concept that such striking progress has been made, certainly in Britain, in the realization of so many of these objectives.

The Benefits Achieved

Not more than six years after these points were made, it was found possible to hold, in Britain, Quality and Reliability Year. The results of this national effort have been published.[4] Its success was due, more than anything else, to the combination of 'reliability' with 'quality'. Quality, after all, is sometimes difficult to define, but everybody appreciates reliability. It is a short step to the involvement of the supplier (who may be a machine operator) with his customer (who may be himself buying the car for which he has made some minor component, or as a passenger on an airline depending upon some part for which he has been responsible).

The results achieved in some of the larger British companies have confirmed beyond doubt the substantial savings which can be made by an objective approach to quality achievements. In other cases, overwhelmingly successful overseas sales have been achieved, mainly by reason of the quality of design and the reputation for reliability of the firms concerned. But in addition, and to the pleasurable surprise of the managements involved, it has been shown that Q & R, about the merits of which there can be no argument, has turned out to be the finest means of achieving the involvement of all employees in the end-purpose of the company—item (n) of the points listed above.

This chapter began by instancing the case of the Production Director who thought that quality might increase his costs.

24

From the examples given it is clear that nobody, no matter what his function, can afford to ignore the importance of the quality and reliability of his contribution towards the end of satisfying his customer.

References

1. Cleaver, P. C. 'The cost of unreliability to the Royal Air Force.' *Proceedings, Society of Environmental Engineers,* 1963.
2. Hedger, E. F. 'Incentive contracting.' *The· Reliability of Service Equipment,* I.Mech.E., 1968.
3. Nixon, F. 'The value of the control of product quality.' Conference of the European Organization for Quality Control, London, 1960.
4. 'Profiting by quality and reliability.' Report of National Conference 1966 (with 41 Case Studies). National Council for Quality and Reliability.

3
The Evolution of Quality and Reliability

The first task of the executives of an industrial enterprise is to decide which sector of the market they intend to be theirs. Next, they must decide what type of product is most likely to meet the market's needs. Those who realize that to succeed they must satisfy their customers then pose to themselves the questions, 'How can quality and reliability be achieved?'; 'What action must be taken within the organization to ensure that the desired end be achieved?'

For many years past, anyone voicing these questions has run the risk of being deafened by offers of help from people claiming an expert knowledge of either quality or reliability, but seldom of both. Their proposals have usually involved the introduction of statistical techniques into the organization, and more often than not they have failed to convince managers who have an instinctive distrust of some of the arguments put forward.

It is rather sad to consider the wasted years during which industry could have availed itself of the benefits of Q & R. The fundamental approach, established in 1924 by the Bell Telephone Laboratories, became obscured by excessive preoccupation with statistical methods, which were seen originally as no more than useful aids. Fortunately, during the past ten years or so, the British Q & R approach, which is more or less a reversion to the original concept, is proving to be understandable by and therefore acceptable to management. There is indeed evidence that management is showing a keener interest in Q & R in Britain than in any other country. Dr A. V. Feigenbaum[1] has said:

> In the U.K. the quality control function as we know it is only one factor in nation-wide developments. . . . This approach wisely recognises that men with a wide variety of industrial, trade union and governmental backgrounds have as much interest in the quality field as do the quality professionals.

Nevertheless, there remain large numbers of uncommitted managements. It will help them towards a clearer understanding of Q & R, and of the answers to the questions posed earlier, if we trace the evolution

26

and development of the Q & R philosophy from its beginnings. It will be shown how the original concept was basically sound; why and how it was not so widely adopted as it deserved to be, due to confusion between ends and means; how the British Q & R concept is recovering the situation; and how the remanence of misconceptions still provides pitfalls for the unwary.

The Pioneers

The Bell Telephone Laboratories It is established beyond doubt that credit for originating and developing the philosophy and practice of satisfying the customer by assuring the quality of the product must go to the late Dr R. L. Jones and his colleagues of the Bell Telephone Laboratories.

They began their work as a small team in the Western Electric. Company—at the well-known Hawthorne plant—which is the manufacturing unit of the Bell System. With production even in those days running at a rate of about 10 million telephone handsets annually, and rising rapidly, studies were initiated to discover means of ensuring maximum economic output of satisfactory product. In 1924, Jones was put in charge of a new Inspection Engineering Department. In the following year, this department was incorporated in the Bell Telephone Laboratories and soon afterwards its title was changed to Quality Assurance.

Jones saw the function of his department to be the optimization of output of reliable product. He built up a team comprising 'men of many talents: physicists, mathematicians, experts in glass, metals, wood, leather, paper, textiles; engineers experienced in design, manufacture and operation'. His description of the duties of his department bears a close resemblance to that which is applicable to the more advanced practices of today. These duties were, briefly:

(a) To develop the theory of inspection, statistical methods, and new principles.
(b) To develop methods of specifying the quality and establishing economic standards of quality of telephone equipment.
(c) To maintain oversight of the quality of outgoing goods.
(d) To study the performance of equipment in service and to guide the steps taken to prevent recurrence of trouble.

In retrospect, it is little short of tragic that Jones' pioneer work received scant notice, mainly because it was described only in the *Bell Laboratories Record*[2] which, as a house magazine, did not have a wide external circulation.

The team which Jones established consisted of young engineers and scientists, most of whom have played important parts in the subsequent development of quality control. The names of G. D. Edwards, D. A. Quarles, H. F. Dodge, E. G. D. Paterson, W. A. Shewhart, J. M. Juran, H. G. Romig, and Miss M. N. Torrey will be familiar to all who have been interested in statistics and quality control. The Bell approach, however, was and is much more broadly-based than has usually been appreciated by many of those who have endeavoured to emulate the pioneers. For example, Jones quite clearly understood that reliability must be a main aim, and that service experience and optimization of the design for maximum value were important factors.

According to R. B. Murphy, to whom, as well as to E. G. D. Paterson, the author is indebted for interesting historical detail,[3] the late C. N.

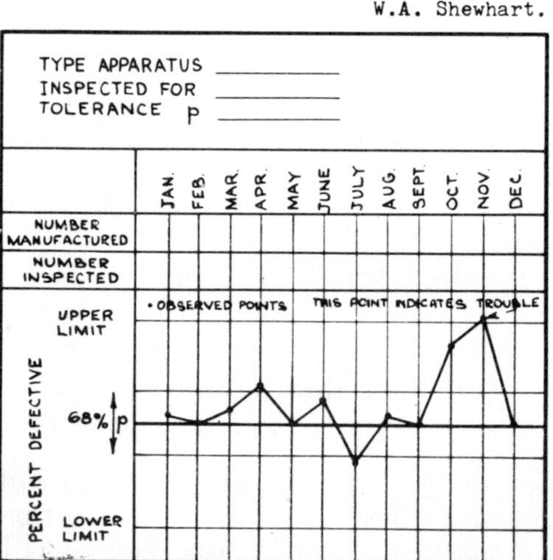

Case 18013   WAS-724-5-16-24-FQ

Mr. R.L. JONES:-

A few days ago, you mentioned some of the problems connected with the development of an acceptable form of inspection report which might be modified from time to time, in order to give at a glance the greatest amount of accurate information.

The attached form of report is designed to indicate whether or not the observed variations in the percent of defective apparatus of a given type are significant; that is, to indicate whether or not the product is satisfactory. The theory underlying the method of determining the significance of the variations in the value of p is somewhat involved when considered in such a form as to cover practically all types of problems. I have already started the preparation of a series of memoranda covering these points in detail. Should it be found desirable, however, to make use of this form of chart in any of the studies now being conducted within the Inspection Department, it will be possible to indicate the method to be followed in the particular examples.

W.A. Shewhart.

Enc.:
Form of Report

Fig. 3.1. Shewart's Historic Announcement of the Control Chart (16 May 1924).

Frazee of the Western Electric Company was probably the first to realize that the statistical sampling methods already being used in scientific fields could be applied to industrial processes. As the recruit most recently arrived from university, W. A. Shewhart, a physicist, was directed to study the potential value of these methods. On 16 May 1924 he wrote to R. L. Jones informing him of his control chart concept (Fig. 3.1). Subsequently, methods of determining the probable quality of batches of product by inspecting small samples were developed by Dodge and Romig.

In 1931, Shewhart's book *Economic Control of Quality of Manufactured Product* was published. It was to have a great and unanticipated influence throughout the world. In it Shewhart described the applications of the statistical methods which had been evolved. It can safely be assumed that he was entirely familiar with the basic principles which had been laid down by Jones. Apparently, he assumed that his readers would be equally informed. Owing to the limited circulation of Jones' paper this was not the case, and unfortunately many people accepted the methods described by Shewhart as an end in themselves. Shewhart made no such mistake, as he made clear in the following statement.

> The control of quality of manufactured product involves three co-ordinate functional steps: the specification of the aimed-at standard of quality; the production of pieces of product that will be of standard quality; and the determination of whether or not product thus made is of standard quality.[4]

It would appear that many of those who read Shewhart's book failed to realize these fundamentals, being more interested in the new statistical tools which he was describing. This failure to appreciate the fundamentals, and the overenthusiastic espousal of statistical techniques, have deprived the industry of many countries of years of productive performance at higher levels of efficiency.

Early Developments in Britain In 1932 Shewhart visited Britain. Here, Dr B. P. Dudding of the General Electric Company Ltd had already been applying statistical methods to the study of the quality of electric lamps. L. H. C. Tippett had begun to apply statistics to the problems of textile manufacture, and Professor Egon S. Pearson provided a direct link with his father Professor Karl Pearson and with W. S. Gossett (who wrote under the pseudonym 'Student'), who had applied statistical sampling to the improvement of strains of barley and brews of Guinness' stout nearly 30 years earlier.

Of greater fundamental importance, although it has long gone unrecognized, was the contribution made over half a century ago by the aeronautical side of the British War Office. In 1909 Mervyn O'Gorman, a professional engineer, was appointed Superintendent of the Balloon Factory at Farnborough. Here were engineers, civil and military, working on aeroplanes. These men saw the need to establish design and manufacture upon a formal basis and in 1912 the first Airworthiness Certificate was granted, to a BEI aircraft, on completion of satisfactory static load tests on the structure.

Early in 1914 the Aeronautical Inspection Department of the Royal Flying Corps was established under Major J. D. B. Fulton, RFA. There was at this time a lively aircraft design and manufacturing industry, but no British manufacture of aero engines.

Still before the outbreak of war, C. G. Grey was reporting in *The Aeroplane* that the AID were 'getting into order a proper system of what one must, I suppose, call "book-keeping", by which every incident can be traced.'

By 1915 there was a large production of aircraft and engines in Britain. The AID played an important part in training the new industry. AID approval became a hallmark of quality, and Lord Weir stated later that the AID had 'played a definite part in raising the level of British engineering quality'. Nevertheless, by 1917 the AID was being criticized for being too rigid in its interpretation of requirements—due to dilution and a lowering of the standard of the rank and file inspectors.

Fulton died in 1915 and his place was taken eventually by Lt. Col. H. W. S. Outram, who also died in harness, as Director-General, in 1944.

Thus, there was early insistence upon conformance of product and process with specified requirements. Records had to be kept, and all changes, to assemblies, detail drawings, processes and standards had to be approved by the customer's representative—the resident AID Inspector-in-Charge.

Thus, although the phrase was not used until Bell adopted it about 10 years later, the basic requirements of quality assurance were laid down. The system has in fact become so much the way of life of British aircraft firms that it is sometimes difficult to recognize that present-day efforts, American and British, to introduce similar practices into defence industries are really nothing new.

Additional confirmation of the soundness of the system is provided by automobile manufacturers. To comply with the US National Traffic and Motor Car Safety Act 1966, they have found it necessary to introduce a system of design, manufacturing, and quality control which is practically identical with that of the early AID-trained aircraft firms.

It is a great pity that those who were charged with effecting economies in Government expenditure in the years following the Second World War failed to understand that 'inspection' was not the major role of the AID. Twenty five years elapsed before the importance of the underlying system of quality assurance was properly appreciated.

Opportunities Missed

Even in the United States, industry was at first a little slow to follow the lead of Jones and his team at the Bell Telephone Laboratories.

The Second World War provided an incentive. Unfortunately, the circumstances of greatly increased rates of production lent themselves readily to the application of the new statistical techniques, and the fundamentals of quality assurance were largely ignored. Intensive campaigns were conducted across the United States, to train thousands of people in the techniques of *statistical quality control*. The three essential requirements outlined by Shewhart were appreciated by only a few. Large numbers of SQC practitioners found work in machine shops, and it probably did not occur to them that the detail drawing, with its dimensions and tolerances, was in fact the specification which had to be met.

In both the United States and Britain the education campaigns were Government-sponsored. Whereas in the United States the task was delegated mainly to the universities, in Britain Government participation was the more obvious, the chosen instrument being a Government statistical advisory unit. Possibly because those who helped in the American campaign had come under the direct influence of the enthusiastic and experienced Bell team, their work had a more permanent effect. In Britain, quality control seems to have been adopted rather more from compulsion than from conviction. Hence, when Government controls were removed at the end of the war, quality *control* was an early victim. The fact that after years of shortages there was a seller's market meant too that the incentive of competition was absent.

Other factors influenced the situation in Britain. As Tippett has said:

> Statisticians like myself are apt to call the subject statistical quality control, hoping thereby to absolve themselves from the duty of considering management aspects. [5]

In consequence, there was little understanding of basic principles. A common sight in British factories during the war was row upon row of control charts mounted in expensive metal frames, one to each machine, standing in dusty futility. There is little point in maintaining control charts on processes which are consistently in control, as the majority of machining processes usually are. Management and operatives alike became quickly disillusioned as to their value.

It was unfortunate, too, that Professor Egon Pearson's excellent guide[6] went out of print just before the war. A second impression was planned, but this intention was frustrated when the set type was destroyed by enemy action.

Whatever the reasons, the fact remains that in the United States there was earlier a better and more understanding appreciation, by management, of the fundamentals of quality control. This was accompanied by a rapid growth of professionalism, so rapid in fact that today the membership of the American Society for Quality Control is in excess of 20,000. It is perhaps indicative of the pattern of developments in the United States during the past 30 years, however, that it is only within the past two or three years that the ASQC has set up Reliability and Inspection Divisions.

During the postwar years statistical quality control became well-established, on the continent and in Britain, in firms employing continuous processes, as in the chemical industry and papermaking. The large quantity production units of the rapidly growing electrical and electronic industries followed. In Holland particularly, the home of Philips and Unilever, statistics gained a firm hold. In Britain, with its large numbers of old-established engineering firms, with much of its industry concerned with short runs and small batch production, the applications of statistical methods have been more restricted, as already indicated. Indeed, there can be little doubt that the overemphasis upon statistical techniques, in areas where their value was limited, retarded the adoption of quality control.

This may well prove to have been an advantage in the long run. This is probably because British reluctance to be stampeded into acceptance of the new techniques led to a thorough study of the fundamental factors underlying the achievement of product quality and reliability.

A Return to Fundamentals

During the 'fifties the managements of Britain, the United States, and Japan began to take an increasing interest in the advantages which might accrue from the adoption of quality and reliability principles. These movements had different motivations, but although they have followed different paths they are tending towards a common pattern which resembles closely the original concept of Bell.

In Britain, the British Productivity Council had been struggling to spread an interest in quality control. Independently, the Institution of

Production Engineers set up a committee which prepared for publication a valuable booklet which showed a new and refreshing appreciation of the basic approach.[7] It was certainly the first British publication to emphasize the responsibility of senior management, and to make clear the necessity to involve all departments concerned with the design and manufacture of the product.

About 10 years ago a study was made of the reasons for the difference between American and British progress since the war. It became apparent that many British industrialists had an innate suspicion of the predominantly statistical approach to quality control which then prevailed. It was clear, too, that the conflict between the quality and reliability interests in the United States, which still exists to some extent, was benefiting neither party. It was sensed instinctively that quality and reliability had always been implicit but undefined factors of great importance in the planning and direction of successful British enterprises. Research among consumers, and more particularly among housewives, showed that most of them looked upon 'quality' as a desirable attribute, but one which was beyond the reach of their purse. They thought of quality in its sense of 'degree of excellence', so that a quality product meant the best obtainable.

They experienced no such difficulty when it came to considering reliability. A common definition of what was really desired by the individual customer was 'a product which went on working properly for as long as it was felt that it ought to, for the money paid for it'. Add the proviso that the user must treat the product fairly, and not abuse it, and we have a far better definition of reliability than some of those evolved for military weaponry.

At this time, the suggestion was being made that there should be set up a specialist institution of reliability engineers in Britain. It was realized, however, that such a move would probably retard rather than advance progress towards the wider involvement of managers and engineers throughout industry which was felt to be essential.

At the Fourth Annual Conference of the European Organization for Quality Control, which was held in London in 1960, my presidential address included the following:

> The Purpose of an Industrial Organization is to Achieve the Requisite Quality of Product, at Minimum Cost.
>
> Requisite quality is defined as that which is necessary to ensure customer satisfaction, by achieving a given standard of reliability, or fitness for purpose, in the product.
>
> The exercise of full control of product quality will require the

application of commonsense, engineering, inspection, proving (e.g. by testing), and statistical methods of analysing manufacturing performance, each in greater or lesser degree according to the nature of the product, its value, its purpose, the rate and scale of its production, the personality of the organization, and the capability of the personnel concerned. . . .

There is no simple universal solution, nor any single ideal form of organization. The precise title under which each of the basic acts is operated is of no significance, and it may well vary in different organizations.

The National Council for Quality and Reliability In 1961, as a result of the active encouragement of the British Productivity Council and of individual members of other bodies, notably the Institution of Production Engineers, the National Council for Quality and Reliability was formed.

Its constitution was unusual. It was set up to promote, throughout the country, an awareness of the importance of achieving quality and reliability in the design, manufacture, and use of British products, and in services throughout the private and public sectors. Its membership consists of 45 national bodies which include many of the senior professional institutions, Government departments in defence and technology, the Trades Union Congress, the British Institute of Management, and the Institute of Directors.

It wields no executive authority, nor can it with such a membership of widely diverse interests have a single voice, save on matters of broadest principle. It functions by exercising an instigatory and catalytic influence upon its members. Many of them have initiated or sponsored conferences on a national scale, some of which have been of great significance. Since the aggregate membership of its member organizations is of the order of 750,000, not counting the millions of trade union members and tens of thousands of civil servants, the potential influence of NCQR greatly exceeds that of any other body concerned with quality and reliability.

In 1966–67, only five years after the formation of NCQR, it was found possible for the British Productivity Council, its sponsor, to launch Quality and Reliability Year. Upwards of 8000 industrial concerns participated, and more than 1400 of them held campaigns within their organizations. The campaign was noteworthy for the extent to which it succeeded in involving senior management. In fact, this was a major factor enabling the campaign to be mounted so quickly, completely, and effectively.

34

Management's Task In 1963, I was invited to present a paper at the XIIIth Triennial Conference in New York of the Comité International de l'Organisation Scientifique, on the latest developments in quality and reliability in Europe. There was no doubt that the organizers expected a dissertation upon some new refinement of current techniques. Realizing that Rolls-Royce had enjoyed for nearly 60 years a unique reputation for the quality and reliability of its products, a study was made of how this had been achieved. It soon became apparent that this reputation had been established long before quality and reliability had become catch-words in many firms, cults in others. Further investigation showed that the company, at that time only about three years old and employing only a few hundred people, had an understanding of organizational requirements and objectives which, though rare today, was by no means unusual in those days, before management had become confused by today's plethora of techniques. The company's continuing success has been due in large measure to its retention and observation of the principles upon which its reputation was founded.

It was possible to identify the following:

Essential Requirements for the Achievement of Reliable Product[8]

(a) A satisfactory design of product, thoroughly proved by adequate testing in order to establish its reliability under the conditions to which it will be subjected in use.

(b) A full specification of the requirements of the design, which must be clearly understood by everyone concerned with the production of the constituent parts and the complete end-product.

(c) Confirmation that the manufacturing processes are capable of meeting these requirements.

(d) Full acceptance, by all those concerned with production, of the responsibility for meeting the standards set by the specification.

(e) A check that the products conform with the specification. This check is required to protect the customer, to safeguard the reputation of the company, and to provide essential information regarding failure to conform.

(f) Instruction in the use of the product.

(g) A study of user experience, feedback to the department concerned, and rapid remedial action.

When offering these essentials at the International Conference on Quality Control, Tokyo 1969, two more were added to the list:

(h) Everyone in the organization 'working till it hurts', in the words of the late Lord Hives.

(i) Realization that the customer must receive better satisfaction than he can obtain from the company's competitors.

The foregoing is merely a statement of the obvious. Nevertheless, contact with many companies in different industries in many different countries has shown that it is rare to find firms having a clear understanding of these requirements, still rarer to find them all properly implemented. In the majority of cases, it has been found that there exists a greater or lesser degree of neglect in the following areas:

(a) Failure of design to take all service requirements into account.
(b) Too little development testing to reveal potential deficiencies in the design.
(c) Little or no effort to educate the user.
(d) Neglect of experience in the field.

The essentials listed above show clearly the extent to which management must involve departments throughout the organization in the effort necessary to achieve reliability. Indeed, only senior management can ensure the direction and coordination which are obviously needed.

Meanwhile, in the United States, there had appeared, in 1951, Dr J. M. Juran's *Quality Control Handbook*, and Dr A. V. Feigenbaum's book *Quality Control*. Both of these works emphasized the role of management.

At about the same time Dr W. Edwards Deming had begun a series of visits to Japan. Here, he played an important part in showing Japanese executives how to incorporate the quality philosophy in the organizations of the companies which they were re-establishing after the war. At first, statistical techniques were taken up with great enthusiasm. Courses of instruction were given on radio, on television, and in the factories. Thanks to Dr Deming's clear exposition of fundamentals, however, Japanese firms were probably even quicker than those in the United States to see clearly the vital part which management must play.

> Companies that have promoted it as 'Q.C. equals management control' are very prosperous, while those that have conducted Q.C. within a small part of the inspection department by interpreting Q.C. in its narrow sense have not been so successful.[9]

Recognizing the general movement towards greater involvement of management, Feigenbaum produced, in 1961, a new book *Total Quality Control*. This was followed in 1963 by a second edition of Juran's *Handbook*, greatly enlarged, and in 1964 by his *Managerial Breakthrough*.

Perhaps the greatest influence has been exercised, however, by the US Department of Defense, which in 1959 published Specification

36

MIL-Q-9858 'Quality Program Requirements'. Its successor, MIL-Q-9858A, published in 1963, has made it incumbent upon contractors to the Department to have within their organization a system of control which is similar to that established in Britain's aircraft industry 50 years ago.

Another development the full effect of which is not fully appreciated is the American National Traffic and Motor Car Safety Act, 1966. For their own protection, as well as that of motorists and pedestrians alike, main and ancillary firms in the automobile industries of every nation must, sooner or later, adopt similar systems.

Some Pitfalls

Simple, clear, and obvious as may be the principles outlined by Jones' team at the Bell Telephone Laboratories, there remain many who are still not fully aware of them. These people tend to fall easy prey to those who profess to an understanding which in fact they do not possess.

Regrettably, considerable confusion still exists, in many circles, about the real purpose and function of various activities connected with the overall aim of quality assurance. On the whole, it is people who feel that their jobs are threatened—inspectors who fear that quality control will pass them by—or aspirants who do not understand the management viewpoint, who have created the confusion. One symptom of it is the number of different titles which are used in different organizations to describe functions which are similar, and often identical. Inspection, Quality Control, Statistical Quality Control, Quality Audit, and Quality Assurance describe what really are different activities. Yet they are often applied, indiscriminately and with little understanding, to what is often no more than post-mortem inspection. Quality Engineering and Total Quality Control are other titles which do little to simplify the situation. Implicit in many of them still remains the suggestion that statistical methods play a vital part.

In much the same way, reliability is interpreted in entirely different ways by people concerned with missiles and electronic and nuclear equipments on the one hand, and with mechanical products like aero engines and motor cars and lawn mowers on the other. The former, who were extremely concerned from 10 to 12 years ago by the then unacceptably high unreliability of highly complicated products, 'went statistical'. They had some justification for doing so, because the inherent unreliability, by mechanical engineering standards, and the large number of components involved, provided a sound enough basis for the statistical probability approach.

On the other hand, under the all-pervading compulsion of safety of life, engineers and managers in all branches of the aircraft industry have always had a clear understanding of the fundamentals of reliability achievement. The principles and practices which they follow are applicable to all types of industry, yet it is only recently that they are beginning to be adopted.

It has been interesting to see how the challenges posed first by nuclear submarines, and later by the American National Traffic and Motor Car Safety Act 1966, have led to the introduction of aircraft engineering and management practices by shipbuilders and motor car manufacturers. During recent years, there has been a great improvement in the reliability of detail electronic components. This means that the statistical sampling methods used earlier are diminishing in value, for the reasons which apply to mechanical equipment. The fewer components of a complex device that are likely to be unreliable, the larger the test sample that is needed. Hence in this field, too, we have seen the adoption of engineering and management systems which derive from the aircraft industry. There can be little doubt that this has been a factor in the success of the Apollo expeditions to the moon.

References

1. 'Quality programs for the 1970's.' *Quality Progress*, December 1968.
2. Jones, R. L. 'The viewpoint of inspection engineering.' *Bell Laboratories Record*. Vol. II, No. 6, August 1926.
3. Paterson, E. G. D. Papers reproduced in *Industrial Quality Control*, May, July, August 1960.
4. Shewhart, W. A. 'Nature and origin of standards of quality.' *Bell System Technical Journal*, No. 1, Vol. XXXVII, January 1958, reproducing a paper dated 1935.
5. Tippett, L. H. C. *Technological Applications of Statistics*. Williams and Norgate Ltd., 1952, London.
6. Pearson, E. S. 'The Application of Statistical Methods to Industrial Standardization and Quality Control.' BS 600, British Standards Institution, London, 1935.
7. *Quality: Its Creation and Control*. Institution of Production Engineers, 1958, London. Reissued 1962.
8. Nixon, F. 'Quality management in Europe's expanding markets.' CIOS XIII, 1963, New York.
9. 'Quality Control in Japan.' Report of Quality Control Specialists Study Team, January 1958, Tokyo.

Industrial Enterprise

4

The Manufacturing Complex

Since the days of Charles Babbage so much has been written about business organization, manufacturing, and productivity that a further essay on these subjects could only be repetitious and boring. There is, however, one aspect of industrial enterprise which has so far received too little notice. It concerns the number of interdependent firms involved in each separate project.

> Even the smallest concern, the village shoemaker for example, must buy leather from tanners, who buy skins, tannic acid, and dressings from other companies. The shoemaker must obtain wax and thread, nails and eyelets from specialist suppliers, and he will use a modicum of machinery. Needles will be a consumable item of equipment, and once in a while he will require a new apron, rasps, a hammer and a last, as these become damaged or wear out.

As manufacturing organizations become larger, their dependence upon outside sources of supply, for materials, equipment, and services increases rapidly. A firm producing motor cars, or accessories for motor cars, will probably have as many as 700 to 1000 main suppliers. They will be responsible for up to 70 per cent of the value of the end product. They will include suppliers of materials, steel and alloy bar, sheet, forgings and castings; manufacturers of standard detail components such as nuts, bolts, ball and roller bearings; makers of proprietary products, for example pumps, carburettors, brake and steering assemblies, gearboxes, locks, and handles. Specialist manufacturers will supply pistons, piston rings, valves, and valve springs; other companies will provide machining services, making parts under sub-contract to the primary firm's own drawings. These subsidiary firms will themselves depend upon perhaps several hundred other suppliers. This same pattern applies throughout industry. The only difference is that somewhat fewer firms might be involved when the end-product is less complicated than a motor car.

At the other extreme, some large firms in the American aerospace industry have upwards of 7000 suppliers. The US Department of Defense has 18,000 contractors, and the mind boggles at the total number of enterprises involved, directly and indirectly, in supplying the Department of Defense with equipment. The majority of defence agencies, and large retailers such as Marks and Spencer Ltd, make nothing themselves. They are entirely dependent upon outside suppliers.

In the ultimate, it is probably no exaggeration to say that every manufacturing organization in the country bears some responsibility, however indirect or however small, to every primary product. In addition, every organization depends upon people. A company's employees are indeed among its suppliers. Whether it employs ten or a hundred thousand people, the firm is customer to all these individual men and women who supply their knowledge, intelligence, and skills in exchange for wages and salaries, job interest, and fulfilment.

The Challenge to Management
This complex situation brings in its train a host of difficulties which are at the root of most of the troubles encountered in industrial relationships.

Every Production Manager has suffered the frustrations caused by schedules of output which have been ruined by the late arrival of essential items, or by a batch of material or parts which is found to be unsatisfactory on receipt. Or it may be the late arrival of an important machine tool, or the breakdown of another one, and non-availability of a replacement part which makes it impossible to meet the desired programme.

It is doubtful whether people responsible for late deliveries of supplies, or poor reliability of equipment, are aware of their failure. The supplier of steel bar or billet, for example, cannot easily visualize the form which vital components made from that material will eventually take. Nor will he understand, generally, the conditions of stress and the importance of the quality of what are to him merely lumps of metal. The long and complicated cycle of conversion, the remoteness both in time and technical complexity, form a barrier to the realization and acceptance of responsibility for the proper performance of that metal.

What other reasons can there be for the reluctance of steel suppliers, for example, to guarantee the quality of their steel? And for their long established resistance to suggestions that steel specifications should be quantified? It would be a healthy exercise for material suppliers to consider how they themselves would react should they be compelled to buy furnaces, machinery, or their own motor cars, under the conditions of sale which they offer to their customers (and impose upon them, when there is a seller's market).

If management wishes to bring a major producing company to the point of maximum efficiency, all individuals and firms supporting the enterprise must identify themselves with the end-purpose. Yet only a minority of managements appreciate this. A study of the extent to which a company is dependent upon so many people, within and outside its direct control, can so broaden the viewpoint of that company's senior management that by taking appropriate action it must inevitably benefit.

That this is not better appreciated is due probably to the discipline to which most managers are subjected. They are conditioned to achieve

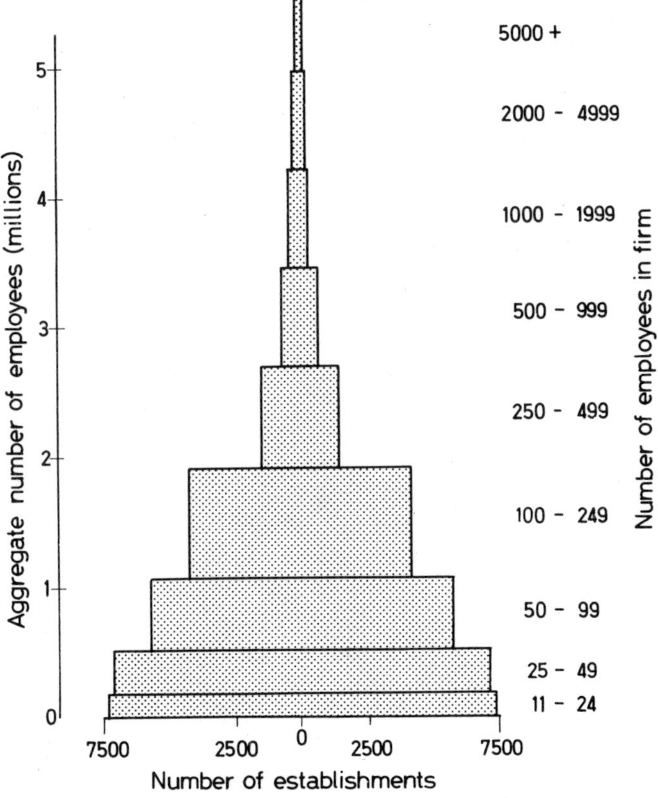

Fig. 4.1. The Composition of British Industry.

the short-term or immediate result. The prospect of a long-term effort to effect long-term benefits is itself something of a deterrent, since it usually interferes with the immediate task. Yet the advantages and benefits are considerable. By making supplying firms realize their responsibilities not only can promised deliveries be maintained, product quality improved, lead times shortened, and stockholdings reduced, but also there will be an overall increase in efficiency and a consequent reduction in costs.

Leading firms are already taking action to involve their suppliers in their own end-purpose. Each step taken has an ever-widening effect. By acting as a tonic to each group of suppliers, it influences their own suppliers in turn. In consequence, several thousand firms may be influenced for the better as the result of the action of one company. During National Quality and Reliability Year (1966–67), this movement towards total involvement, which had already been started by a few firms, received further impetus.

The definition and allocation of responsibility throughout the whole of the complex involved in the effort to produce an equipment or a service is in fact a specific requirement which contractors to the US Department of Defense and the British Navy Department of the Ministry of Defence must now meet.

Figures 4.1 and 4.2 show the composition of British and American industry, in terms of the number of firms of each size, large and small. These analyses ignore the smallest firms, the tens of thousands employing fewer than ten employees. Yet these small firms are not negligible to the efforts of the large industrial empires. It is a fact that some of the largest firms in Britain obtain essential supplies from some of the smallest.

In addition, each and every one of these firms depends upon people. A company's employees are one of its major assets. The involvement of its employees in the purpose of the enterprise is one of the prime responsibilities of management. Q & R provides one of the best means which has yet been evolved for the motivation of people. Because every man can see clearly that quality and reliability are good things in his own individual case, he can appreciate that they are good for everyone else as well. It has been found that bringing this to the attention of its employees will of itself ensure involvement in a company's aims.

It is useful to remember, too, that a company's employees are in their turn the customers of suppliers of food and clothing, shelter and public transport, amusement and recreation, protection (as by police and the armed services), medical care, and organization and coordination (as by local and national Government). Without this interdependent

44

complex of primary suppliers with the other supplying agencies, there could be no industry.

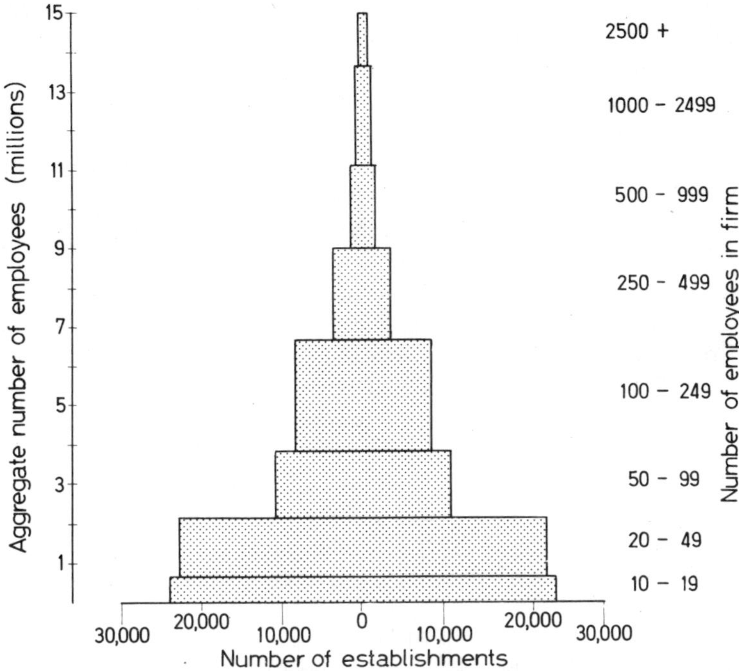

Fig. 4.2. The Composition of American Industry.

Looking Both Ways

In every case, whether we are considering the individual employee or the industrial organization, there is a dual attitude of mind. We are, each one of us, at one and the same time both a supplier and a customer. Indeed, in not a few cases a supplier may be the customer of his own customer. A man may buy a bicycle, a television set, a motor car, or bread which he has himself helped to make. A supplier of castings or of ball bearings may buy a machine tool embodying components of his own manufacture. The question should be posed—is he entitled to receive better service and value than he himself provides?

Consideration of the dual role of supplier and customer enables management to see more clearly *what* must be done to achieve the requisite quality and reliability of product, and *who* must do *what*.

In the general case, unfortunately, we are more conscious of our rights, of our personal dissatisfactions as customers than of our duties as suppliers. Some large supplying firms, however, have clearly realized their responsibilities to their customers. Firms which have done this usually prove to be good customers in their turn. They have come to expect cooperation and assistance from those whom they supply, and they realize that they themselves can benefit by offering similar help to those who supply them. They have appreciated that for them to offer optimum value to their customers it is necessary for them to enlist the maximum support from their own suppliers, whose end-customer is not themselves, but their own customers. They make it clear to all the parties concerned that should they not succeed in holding their market by providing customer satisfaction, they will cease to be a market to their present suppliers. In this interdependent cycle, they are agents as well as doers.

To achieve the involvement of a firm's suppliers in the total enterprise of the firm requires in the first instance a recognition by senior management of the importance of this involvement. This must be followed by action, by the same senior management, and through the medium of the purchasing department, to enlist the cooperation and support of all the suppliers concerned.

This holds good as much when we are considering the individual employee as a supplier of services to his firm as when the supplier is a company, small or large. In every case, it is incumbent upon the customer (or the employer or his agent, the manager) to make certain that his supplier (or employee):

(a) Knows what is expected of him.
(b) Has the ability, the will, and the facilities to do what is expected of him.
(c) Carries out his duties satisfactorily.
(d) Can be seen to have done so.

Business is a Two-Party Affair
At each of the many interfaces in the complex of supplier/customer relationships only two parties are involved. As we have seen, each party has a share of the responsibility towards the successful provision of customer satisfaction. These responsibilities will only be met to the full when each party realizes that it stands to benefit from cooperation with the other.

During the past few years, there has been a tendency on the part of the Governments or Government agencies of some countries to presume to ensure customer satisfaction by carrying out independent checks of the quality of suppliers' goods, by examining and approving their 'quality assurance' (*sic*) or inspection organizations.

To a considerable extent these are seen to be misguided attempts by inspectorates and/or standards organizations to exercise 'power'. In the words of Dr Juran,[1] such bodies are aggressive. Often, they fail to understand the fundamentals of business life. Let us examine, however, when and when not the intrusion of a third party into the two-party interface is justifiable and desirable.

The Intrusion of a Third Party

There are certain circumstances in which a third party may justifiably take action, as when there is Government legislation to lay down standards which must be maintained in order to protect the health or safety of the people. Unfortunately the basis of this third party interest is not always clearly understood, and there are well-meaning but mistaken attempts, by individuals in this country and abroad, to impose standards of quality upon ordinary commercial products, and to 'approve' the quality assurance organizations of the firms producing them.

It can perhaps be taken as a compliment to the efforts of the protagonists of quality and reliability that Government departments in more than one country should have been so impressed by the value of quality assurance that they should wish to make its use mandatory throughout industry. Some go further, and wish to assume the right to certify the effectiveness of the quality assurance systems of industrial firms, and to lay down standards of quality to be reached by their products. This action has been proposed with the expressed intention of increasing the attractiveness and competitiveness of products, especially in foreign markets. In support of this proposal is adduced the example of Japan, in the early days of the recovery of whose industry after the war quality control and the Export Inspection Law played important parts.

In general, this trend is indicative of the failure of these departments to understand the fundamental principle of two-party responsibility, and the relevance of product quality to the customer's needs.

In the cases already discussed, in which a Government department is the customer, there may sometimes be a tendency to look upon the appropriate Inspectorate as a third, independent party. In fact the Inspectorate is acting as the agent of the customer, who in the ultimate is the sailor, the soldier, the airman. The extent to which the Inspectorate may be compelled to apply its own standards is a measure of the ineffectiveness of the specification, or of the lack of standards laid down by the design and procurement authorities.

Justifiable Action In the purely commercial field, there are cases where it is right and proper that the Government, or its approved agents, should lay down standards of cleanliness, of purity, of safety, for the protection of the people.

We are happy when we buy a pint of beer to know that the mark etched on the glass indicates that the right quantity has been drawn for us, and that inspectors have checked the accuracy of beer metering pumps. When we buy a pound (or half a kilogram) of tobacco, or sugar, or butter, we are glad to know that the weights and scales have been checked by an Inspector of Weights and Measures, and that we can confidently expect to receive not less than 1 lb (or 0·5 kg) weight of the product.

With the rapidly increasing amount of electrical domestic equipment in our homes, we are reassured to know that standards have been laid down to ensure the safety of cables, plugs and sockets, of electric irons, electric razors, refrigerators, washing machines, etc.

With food and drugs, standards of cleanliness and purity have been stipulated in the Food and Drugs Act 1955, and they are monitored by inspectors. The safety of civil aircraft is established by the Air Registration Board, of ships by the requirements of the Merchant Shipping Acts. In 1966, there was passed by the US Congress the National Traffic and Motor Vehicle Safety Act, which is being strictly enforced and which is having a big influence upon exporters of motor cars from Britain.

Lloyd's Register of Shipping has a somewhat different basis. It came into being to lay down and to check standards of safety of ships so that the insurers, members of the Society of Lloyds, could with confidence underwrite the risks of loss to the ship-owners.

In all these cases, there is a minimum standard which must be met. There is no attempt to impose upon the industrial firm a pattern of quality assurance organization. All that is required is that those minimum requirements shall be met, and shall be seen to have been met.

The action in Japan, where samples of some types of product are inspected before being passed for export, was justified in the circumstances pertaining in that country after the war. Many industries had been destroyed by bombing. In starting up afresh it was obviously in the common interest to have a central testing facility, until each firm in a particular industry had established itself and could do its own testing. At the same time, there was a traditional image of cheap and shoddy imitative goods. It was essential to live this down if an export trade was to be built up.

Concurrently, Japanese industrialists had enthusiastically espoused the quality cause, under the helpful lead of Dr Edwards Deming and other American experts. The Japanese Export Inspection Law was passed, and the JIS (Japanese Inspection Standard) mark was established, in emulation of the BSI kite mark.

Samples of goods ready for export are inspected on a statistical basis, and the batches which are passed, e.g. of cameras, have a gold sticker affixed bearing 'Passed JCII' (Japanese Camera Inspection Institute). It is a fact that the large and successful Japanese camera companies now maintain strict quality control themselves, and it is highly likely that this additional inspection, which in any case is by no means stringent, is no longer necessary. It is also a fact, established by large British retailers of cameras, that the quality of Japanese cameras is not significantly higher than that of those from several other countries. Nor is it likely that the stamp sells many cameras. The great success of Japanese cameras and other products during the past decade is surely due to their advanced design, and their relatively low price, i.e. it is due to their value to the customer.

In Britain, entirely different conditions have prevailed, and British industry has long enjoyed a reputation for the high quality and reliability of its products. It is difficult to understand, therefore, how it can be suggested that national standards of quality of product, and certified conformance with stipulated quality control practices, can increase the attractiveness, the value, the competitiveness of goods which are on offer to customers who must in any case be unknown to the Government. Only the supplier can have the intimate knowledge of his customer's needs, of the duty likely to be imposed, which is such an essential factor—ensuring customer satisfaction.

Fallacious Reasoning The thinking which underlies this tendency to apply controls by a third party is fallacious because:

(a) there is the assumption that assurance of quality of conformance can of itself ensure customer satisfaction (see page 17).
(b) there is the mistaken notion that national standards, which by definition must specify the minimum quality which is satisfactory, can have any significant bearing on competitiveness of products.
(c) it ignores the importance of *design for purpose*, the quality of which can be assessed by nobody but the customer.

Dangers to Industry There are, too, real dangers in this approach. Specifically they are, collectively, the factors which may have some appeal to people in industry who seek protection rather than the bracing wind of competition. The underlying danger is that they could weaken the sense of responsibility which the successful supplier must feel towards his customer. By so doing, they would defeat the intended object, by reducing rather than increasing value and competitiveness.

Reference

1. Juran, Dr J. M. At the International Conference on Quality Control, Tokyo, Japan, 1969.

5

Conducting the Affairs of an Enterprise

We have seen that the achievement of reliable product involves senior management in two major areas of activity. The first may be looked upon as functional, the planning, organization, coordination, and management of the technical departments which have been indicated on page 35. The second is the direction and motivation of all the people involved in the total effort to evolve satisfactory end-product. As has been pointed out by the National Aeronautics and Space Authority in the United States, there must be

> A planned and systematic pattern of all actions necessary to provide adequate confidence that the end-item will meet all specified requirements. [1]

Modern management education takes into account the need to understand the problems of getting the best out of people. There has, perhaps, been insufficient recognition of the number of people involved, of the extent of the dependence of an enterprise upon the employees of vendors and sub-contractors. On the other hand, the narrow specialization of our educational system makes it relatively rare for a manager to have an adequate knowledge of the technical aspects of the enterprise. This is, more probably than not, the reason underlying the failure, common in many firms in many countries, to appreciate the importance of 'planned and systematic management' which is necessary to ensure the efficient coordination of all technical actions. During recent years the American Department of Defense, conscious of the vast sums of money at risk in the procurement of more than 30 billion dollars' worth of equipment annually, has made great efforts to encourage its contractors to realize the importance of systematic, documented planning. This has been followed by an increasing appreciation of the fact that managers benefit by having technical experience, and engineers benefit by having an understanding of the skills of managing.[2]

> The manager has often been compared with the conductor of an orchestra, in the sense that he is a leader. This simile, taken further, can be illuminating.
>
> The orchestra can function only if each member has clear instructions—his score—which he understands. He must have a good instrument, in proper tune, and he must be able to play it. The orchestration comprises the overall plan, which ensures that there is the requisite number of each type of instrument to produce the desired amount of noise of the right quality and at the right time. The conductor trains and motivates and leads his team to perform, as a coordinated unit, so as to produce the desired result.

The success of a good conductor is far more widely acknowledged than is the success of a good manager. Yet apart from the fact that conductors must have an intimate knowledge of the capabilities of each instrument and its player, their task is simpler than that of managers. They know at once if a player or an instrument is at fault, if the scores are not coordinated. The manager of an industrial enterprise often cannot know until much later, and until large sums of money have been lost, should some unit of his enterprise have failed to function in accord with the others, producing a discordant result. When he does realize that the desired aim has not been achieved, he is usually incapable of seeing that the predominant weakness has lain in the lack of clearly defined instructions at every stage of the complex procedure involved in converting a concept into useful product.

Management

In this chapter, we are considering the role of management. This involves the duties of managers, that body of men who carry out the *act* of managing, to which the word 'management' also applies.

It is interesting to note that the American Society of Mechanical Engineers has a definition of management which closely resembles the classic definition of engineering, which was evolved by Thomas Tredgold for the British Institution of Civil Engineers in 1818. Management is 'the art and science of preparing, organizing and directing human effort applied to control the forces and to utilize the materials of nature for the benefit of man'.

The manager is therefore he who, among his other duties, directs the efforts of men. He may be the foreman, responsible for a team of say

52

30 men, or he may be the head of the whole organization. Under whatever title this latter operates, whether he be the chairman of the board of directors, the managing director, the chief executive, the president, or the executive vice-president, he is still a manager. In this case, he manages managers of subdivisions of the entire enterprise. He may be responsible for a small company employing only a few tens of people, or for an industrial empire with tens or hundreds of thousands of employees and hundreds of managers of many different grades of seniority and responsibility.

All of these men must work to the same set of principles. These derive from the scientific method which was identified and promulgated by Francis Bacon 350 years ago. This can be described as *management method* (or hints to conductors).

The following actions must be carried out:

(a) Determine what is required to be done.
(b) Convert this into clear and unequivocal instructions.
(c) Confirm that everyone concerned with their implementation understands the instructions, and knows what he himself is expected to do.
(d) Check that everyone has the knowledge, ability, and will to do the job correctly, training him if necessary.
(e) Provide the equipment and the procedures necessary to enable the work to be done correctly and efficiently.
(f) Check that the job has been done correctly.

These apply to the typing pool and to the production operative, as much as to an executive at any level of seniority.

Obviously the scale of effort will vary according to the task. As far as top management is concerned, in the technical area we have identified the activities which must exist if reliability of functioning product is to be achieved. It is the responsibility of senior management, indeed no one else has the authority, first to determine the amount of effort which is required in each of the departmental activities in order to meet the requirements of the particular product. The work of the different departments must then be coordinated and monitored to achieve the desired result.

The foregoing might be considered as stating the obvious were it not for the fact that there is a remarkably large number of managers, at all levels, who either do not understand the elements of management method, or fail to apply them. One of the commonest indications of this is the widespread failure to provide written instructions of what must be done, and to require written evidence that it has been done. This is the basic system which was introduced into the British aircraft industry half a century ago. It is the present concern of the US Department of Defense and of the National Aeronautics and Space Authority. It is what is being introduced into their organizations, as rapidly as possible, by British and American automobile manufacturers in order to comply with the requirements of the American National Traffic and Motor Car Safety Act, 1966.

These, however, represent only a small fraction of that portion of a nation's industry which produces functioning machinery and equipment. It will be instructive to examine some of the misconceptions which are held by the majority of managers in the remainder of industry, who have insufficient appreciation of what must be done to achieve quality and reliability.

Some Delusions of Managers Contact with hundreds of managers, in many countries, shows that a number of incorrect beliefs are strongly held.

DELUSION 1 'Management by objective' means achieving maximum economy and hence maximum profit to the shareholders.

These aims are admirable provided that they are viewed in the context of the overall aim of the enterprise. Unfortunately, emphasis upon costs and profits, for which managers feel a personal responsibility, tends to exclude other considerations.

Confronting managers with the Q & R concept tends to restore a more balanced perspective.

DELUSION 2 The complaining customer is a nuisance-maker who can justifiably be fobbed off by a stereotyped letter, or at most by the grudging replacement of the defective component.

One large American corporation has estimated that up to forty potential customers hear about each single instance of customer dissatisfaction, and cease to be business prospects.

DELUSION 3 Customer complaints are due to poor workmanship.

With products made in large numbers, such as motor cars, kitchen equipment, and much defence equipment, quality defects have a relatively minor effect upon reliability. An error of design, because it is likely to affect each unit of product, will usually have a much greater effect. Yet the design function is persistently neglected by management.

DELUSION 4 Poor quality is due to careless workmen.

A rough estimate by Dr Juran showed that not more than 20 per cent of quality defects are due to operator error. More detailed investigations suggest that this figure is at the upper limit. In well-planned and tooled

54

machining operations operator error may be as low as 1 to 2 per cent of the total causes of defects. The rest are 'management controllable'.

The managers who criticize the workers are themselves far more often than not directly or indirectly responsible for the causes of manufacturing defects.

DELUSION 5   The chief inspector is responsible for the causes of criticism of defective products by customers.

But how often have standards been laid down for the Inspection Department to work to? Not only is the inspector compelled to lay down his own standards, he has little authority to enforce them. In the absence of agreed standards, the acceptable quality tends to fall as demand increases. In any case, inspection can be no more than a post-mortem activity.

DELUSION 6   The remedy for excessive customer complaints is more inspection.

The fallacy of this should be self-evident. Inspection can only sort good from bad, it cannot prevent defects.

DELUSION 7   Quality control or quality assurance are only new names for the inspection function.

It is a prime purpose of this book to expose this fallacy and to show that quality assurance is the totality of effective management system.

DELUSION 8   The adoption of quality control is all that is necessary to put everything to rights.

Quality control will certainly enable substantial economies to be made. It will have a beneficial but limited effect upon reliability. Quality assurance as defined earlier in this chapter will go much further, but capability of product for the duty to be imposed upon it is of far greater importance, and this depends upon the quality of the design and development engineering.

It is a management responsibility to ensure that adequate effort is devoted to these vital but often neglected activities.

Technique-Oriented Management   The delusions just described are natural to the manager who has been conditioned to look upon the practice of his profession as comprising a collection of specialist techniques. There is no shortage of these. Work study, time and motion study, organization, administration, budgetary control, stock control, motivation, automation, industrial psychology, personnel management, incentives, management by exception, cooperation, involvement, ergonomics, quality control, standard costs, PERT, numerical control, value analysis—you name it, some management consultant will provide it.

A senior member of the Management Board of a large nationalized organization requested information about quality and reliability. Eagerly he asked to be told about 'this new technique'. When he heard that it was largely a matter of commonsense and of managerial responsibility (of which he himself carried a major share) to mobilize, coordinate, and direct the activities of managers and engineers throughout his vast organization, to ensure the exercise of the principles of Q & R to give better and cheaper service to the public, his enthusiasm evaporated.

Like many another senior executive, he had thought that by setting up a department to run a new specialist technique, he would have discharged his responsibility.

No one will gainsay the value of the aids to management which are available. They can all contribute to the overall efficiency of industry. They are no more, however, than facets cut on a rough diamond. When one side of the diamond remains unpolished, when the different facets are not carefully matched to each other in size and direction, the stone will not achieve its full glory. Nor will an industrial enterprise which, by adopting all the appropriate techniques, achieves a high productivity of some product which it cannot sell because its quality and reliability fall short of the market's requirements.

Quality and reliability, of product or service, are the first requirements to be met by the industrial enterprise. Once a market has been won, a wise management will strive to hold it by increasing the product's value to the customer by further improving its quality and reliability, and by reducing costs by the application of the appropriate techniques of management. As the enterprise prospers and grows in size, so it will be found that the specialist techniques will grow in value. Unless the product itself is satisfactory, however, these techniques will be largely cost-ineffective.

It is significant that notable exceptions to the general pattern have been found in firms and organizations facing new technological problems. One which recognized the activities and methods essential to success is one of the earliest recorded—Boulton, Watt and Sons. By 1795, the sons of their better-known fathers were applying the following techniques in their quite small organization:[3]

Market research.*
Planned shop layout.
Production planning.*
Production process standards.*
Machine-operating standards.*

Standardization of components.
Statistical records.*
Detailed cost accounting.
Workers' training.*
Work study.
Payment by results.
Personnel welfare.
Executive development.*
Provision of spare parts.*
Operating instructions.*
Extensive sub-contracting.

The items asterisked are essential to the achievement of product quality and reliability, the rest help to improve the efficiency of the organization.

Only one other of the seventy cases collected by Colonel Urwick as making major contributions to management practice evinced an appreciation of the need for the systematic provision of information. This was Hans Renold, the roller-chain maker. In 1913, he included in a list of the main characteristics of his successful firm 'an organization structure based on a functional specialization and recorded on charts; the existence of written standard management instructions and practices prepared and administered by functional departments; and an interlocking committee system acting as a consultative mechanism among the management staff'.

A contemporary of Renold's was Harrington Emerson. His Principles of Management Efficiency deserve to be reproduced, not only for their soundness, but because they are too little recognized and too little observed:

1. CLEARLY DEFINED IDEAL Know what you are attempting to accomplish. Eliminate vagueness, uncertainty, and aimlessness characteristic of a great many undertakings.
2. COMMON SENSE A supernal common sense that enables one to differentiate between woods and trees. This is a common sense that strives for knowledge and seeks advice from every quarter, unconfined in any position yet maintaining dignity of balance.
3. COMPETENT COUNSEL Actively seeking advice from competent individuals.
4. DISCIPLINE Adherence to rules; strict obedience. The function of this principle is to bring about allegiance to and observance of the remaining eleven principles.

5. FAIR DEAL Justice and fairness.
6. RELIABLE, IMMEDIATE, ADEQUATE, AND PERMANENT RECORDS A call for facts upon which to base decisions.
7. DISPATCHING Scientific planning through which each small function is performed so as to serve to unify the whole and enable the organization to reach its end objective.
8. STANDARDS AND SCHEDULES A method and time for performing tasks.
9. STANDARDIZED CONDITIONS Uniformity of environment.
10. STANDARDIZED OPERATIONS Uniformity of method.
11. WRITTEN STANDARD-PRACTICE INSTRUCTIONS Systematically and accurately reducing practice to writing. (This was Emerson's legal codification for industrial practice.)
12. EFFICIENCY REWARD Reward for successful execution of a given task.

Claude S. George[4] has identified the first five of these principles as applying to problems of personnel (i.e. employer/employee relationships); the remaining seven to 'methodology or systems in management'.

Today, the greatest influence for good, exerting pressure upon industry to adopt the basic principles of Q & R, comes from some of the largest buyers of components and equipment, in widely different fields.

Practically simultaneously two large organizations, one British, one American, realized the vital importance of customer satisfaction, of quality and reliability, and of a clear, mutually understandable and acceptable specification of the requirements of the customer, which must form the basis of a legally enforceable contract between customer and supplier. Each of these organizations was inspired by the need to obtain maximum value for what was, relative to the size of each, a very large expenditure.

The British organization was Marks and Spencer Ltd, the American, the Department of Defense. The former was at the time spending upwards of £100m annually with a large number of suppliers (and in the past few years this figure has been more than doubled). The US Department of Defense's 'buy' is of the order of £12,000m (more than six times the British, and of the order of half the British Gross National Product).

Both organizations realized that acceptance inspection of their purchases would be a costly business, and inefficient because it would be a post-mortem rather than a corrective activity. They realized that the supplier must be held responsible for delivering satisfactory product. To do so, however, he must know precisely what is expected of him.

Marks and Spencer's specification of a shirt, for example, is a model of
its kind. The supplying firms know what is expected of them, and once
it has been established that they can achieve the desired standard, they
are held fully responsible for maintaining it.

The Department of Defense has promulgated its ideas through
MIL-Q-9858A, 'Quality Program Requirements'. This requires of a
contractor to the American Government that he shall have an effective
system, to ensure that all the actions necessary to produce goods which
will meet the specified requirements are fully documented, and that
evidence can be produced, as required, to assure the Government
agency that the work has been carried out correctly. In the words of
John Riordan, Director for Quality and Reliability Assurance in the
Office of the Secretary of Defense, 'the Government does not say how
these subjects are to be tackled and can never directly approve the
contractor's methods, as it would then be sharing in the responsibility
and diminishing the responsibility for delivering products to specification'. The responsibilities referred to are the managerial and engineering
activities necessary to meet the requirements stipulated in the Government contract.

In both examples we see that two parties are involved—the customer,
and the supplier or suppliers. Marks and Spencer Ltd, large customers
themselves, in their turn supply millions of customers, through their
hundreds of retail shops. The Department of Defense acts as the agent
of the armed forces. In the sense that these Services are at the receiving
end of the efforts which have gone before to stipulate a requirement, to
design the equipment, to place orders, and to manufacture the end-
product, the armed forces are customers.

These examples help us to realize that every individual, every department, every organization throughout the whole industrial complex is at
one and the same time a supplier and a customer. Fuller appreciation of
this fact helps towards the clarification of the role of management.

Systematic Management

We have seen that *system* was introduced into the British aircraft industry
half a century ago. Much more recently, its essentiality has been recognized by the American Department of Defense. Management of all
countries should be grateful to this organization for the way in which it
has formalized the requirements, while leaving it to each individual
firm to meet these requirements in the manner best suited to its individual
circumstances.

The Department of Defense's MIL-Q-9858A first appeared in 1959 and it provides the best foundation upon which to build the quality relationship between any two parties.

A few quotations from the specification are illuminating.

> This specification requires the establishment of a quality program by the contractor to assure compliance with the requirements of the contract. The program and procedures used to implement this specification shall be developed by the contractor. The quality program, including procedures, processes and product shall be documented and shall be subject to review by the Government Representative.

> Design of the program shall be based upon consideration of the technical and manufacturing aspects of production and related engineering design and materials. The program shall assure adequate quality throughout all areas of contract performance; for example, design, development, fabrication, processing, assembly, inspection, test, maintenance, packaging, shipping, storage and site installation.

> The program shall provide for the presentation and ready detection of discrepancies and for timely and positive corrective action. The contractor shall make objective evidence of quality conformance readily available to the Government Representative.

> The authority and responsibility of those in charge of the design, production testing, and inspection of quality shall be clearly stated.

> Facilities and standards such as drawings, engineering changes, measuring equipment and the like which are necessary for the creation of the required quality shall be effectively managed. The program shall include an effective control of purchased materials and sub-contracted work.

> Management regularly shall review the status and adequacy of the quality program. The term 'quality program requirements' as used herein identifies the collective requirements of this specification. It does not mean that the fulfillment of the requirements of this specification is the responsibility of any single contractor's organization, function or person.

> . . . all work affecting quality shall be prescribed in clear and completely documented instructions of a type appropriate to the circumstances.

> . . . Records are considered one of the principal forms of objective evidence of quality.

These extracts convey in admirably succinct fashion the basic principles of management responsibility. Several points may be emphasized:

(a) The contractor is made fully responsible for delivering goods in accordance with specification.
(b) The existence of a clear and adequate specification of Government requirements is implicit.
(c) Nowhere does the specification lay down *how* a contractor shall fulfil his obligations.
(d) 'Surveillance' by the Government Representative consists of no more than a review of the contractor's system and of the objective evidence of quality of conformance which must be provided by the contractor.
(e) The functions other than inspection which affect the quality of the product are clearly indicated.
(f) .There is emphasis upon the importance of *system* but none on the organization, the pattern of which is the contractor's business.

An important result of the implementation of this philosophy, according to John Riordan, is that the cost of quality assurance incurred by the Department of Defense has been reduced to 0·47 per cent of the total expenditure on supplies. The enormous costs which would have been incurred had the Department of Defense employed the old-fashioned practice of direct inspection of $30 billion worth of goods provided a powerful incentive to this intelligent delegation of authority.

This specification has provided the basis for a number of others. A few years ago the Navy Department of the British Ministry of Defence was faced with the problems posed by the construction of nuclear-powered submarines. Standards of quality of conformance far higher than ever before were required from shipbuilders who had heretofore followed methods long established by tradition. The American Navy had already coped with the problem successfully, and this experience was drawn upon.

The Ministry of Defence (Navy Department) Specification DPT.200 'General Requirements for the Assurance of Quality (Submarines)' was issued in February 1965. It follows closely the pattern of MIL-Q-9858A. Its implementation, by two different British shipbuilding concerns, resulted in the construction of two nuclear-powered submarines, ahead of schedule and remarkably free from technical snags.

The fact that this revolution in thinking and in practice reached a successful culmination within only half a dozen years is sufficient evidence of its soundness. Other organizations are following this example. The North Atlantic Treaty Military Agency for Standardization published in 1968 'NATO Quality Control System Requirements for Industry (AQAP-1)'. It is an updated equivalent of MIL-Q-9858A, and it has been accepted by countries whose firms tender for the Agency's business. There is a supplement to AQAP-1, 'Guide to the Evaluation of a Contractor's Quality Control System for Compliance with AQAP-1 (AQAP-2)'.[5] This provides a useful primer for the management of any company wishing to set its affairs in up-to-date order.

It is somewhat surprising to find that there are many specifications of requirements issued by organizations other than those mentioned, which fall into the trap of mistaking inspection for quality assurance. With the alleged intention of defining the responsibilities of the contractor, they specify how he shall organize his affairs. By so doing, they diminish the responsibility of the contractor, and provide him with the excuse that if he had not been required to carry out his duties in a certain way, he would have been able to provide a more satisfactory product.

Of course, none of the foregoing can be effective unless it is the subject of a legally-binding contract between the two parties. Before considering the contract it will be useful to examine the actions which must be carried out, the responsibilities which must be accepted, by each of the two parties involved in every business transaction. The two parties are, of course, the customer and the supplier. Since the responsibilities of the customer are too seldom realized, they will be dealt with first.

The Customer's Responsibilities
When the customer is an aggregate of individuals constituting a market for consumer goods, the most that the supplier can expect is that the majority of those people will be sufficiently well-informed as to be able to select the products best suited to their needs, and to treat the products as intended.

As we move on from the individual customer through the small firms in industry to the larger ones and ultimately to the biggest—the industrial giants, the nationalized industries, and services, Government agencies, and Defence Departments—the customer becomes increasingly organized. His responsibilities towards the achievement of reliable product increase in proportion. It will be convenient, therefore, to consider the state of affairs which should exist, the responsibilities which should be accepted by the biggest customers, for example the Defence Departments, collectively buying goods from industry to the annual value of about £600m. The principles apply equally to all large buyers.

In all cases, whether we are considering the Navy, the Army, or the Air Force, the Atomic Energy Authority, a Gas Board or an Electricity Board, or a large industrial concern, there will be a need for new plant or equipment. Its performance and the scale of the requirements will have been determined. In some cases, the design will have been carried out by the authority, in other cases industry will have been invited to submit proposals. Whichever method has been adopted, contracts will be placed with industrial firms for the manufacture of products to the designs laid down.

Clearly, each transaction between the two parties involved should be based upon a full and explicit specification of the requirements of the customer, in which the requisite standards must be laid down. At the same time the customers, in the size category under consideration, have of necessity to place a great deal of dependence upon their suppliers, and they are entitled to require assurance that the money spent will result in the timely delivery of goods of the right quality and reliability.

The ways in which this is done vary widely, and an examination of the methods will lead us to conclusions of value.

Government Departments as Customers Taking first the case of defence equipment, those concerned with the strategic needs of national defence will have determined what is felt to be necessary, in the way of submarines and ships, aircraft and tanks, guns and ammunition, general stores and clothing. We are not concerned here with the way in which the orders are placed with industrial firms, but it can be taken that as a rule this work will not comprise more than a fraction of the total business of the contracting firms. In many cases, the products will be special or, when not novel, will be required to conform to special standards of quality. The special nature of the products will sometimes have required heavy financial investment in plant and equipment, which will have been borne by the Government agency. The defence of the nation will depend upon the delivery of the right goods at the right time, and there will be the need for coordination, in the delivery programmes, of large numbers of different products.

The Government departments are quite justified, therefore, in desiring assurance that their needs will be met. For many years past, this assurance was sought by stationing considerable numbers of Government inspectors in the works of the contractors, ostensibly to oversee the work and to confirm that the products delivered were to specification.

More recently, there has been a growing appreciation of the fact that this was an expensive and not very effective way of assuring quality. There is an increasing tendency to adopt the principles of quality assurance by placing responsibility upon the contractor to deliver goods of the right quality. The customer, the Defence agency, will reserve the right to assure himself that the contractor understands what is required of him, and that he has the necessary system to ensure that the required standards are met.

It may take some time for this approach to be fully understood by both parties. In particular, there is a general weakness in the drawing up of specifications, and a widespread confusion between *organization* and *system*.

For a clear understanding of what is necessary, we must return to fundamentals.

The first step is to determine what is required. In most cases, there will be the need to reconcile conflicting requirements of performance, capability, reliability, price, and delivery date. With defence vehicles—marine, land, or air—there is debate among those requiring speed, or range, or life, or reliability, or maintainability, or size, or offensive power, or environment, or low cost, or early delivery. Not surprisingly, the resolution of these different demands may sometimes take a considerable time, which usually encroaches upon that which should have been available for research, design, and development. All too often a requirement is released to industry before it has been fully agreed and defined. In any case, so much time will have been taken that the importance of laying down clearly defined specifications is usually overlooked, if indeed it were possible to stipulate them in advance of actual experience.

The result of this is that the contractor himself is usually left to decide what standards of quality he thinks appropriate, and he fixes his price accordingly.

Unfortunately, all too often the Government inspectorate concerned will not have been provided with defined and pre-agreed standards. Conscientiously endeavouring to ensure value for the taxpayer's money, it lays down its own standards, and so the seeds of conflict are sown.

Meanwhile, management should bear in mind the differences between the supplier/customer relationship when the customer is a Government agency, and when it is an industrial organization.

The differences will be obvious from the following list of the circumstances which apply, especially in the case of military business, but often too when the customer is a large organization such as a nationalized industry, or a large corporation, or a public utility:

(a) The requirement is determined by the customer, who faces the supplier as a single large (but not necessarily unified) organization.
(b) Often, there is competitive tendering, which usually results in the contract being awarded to the 'lowest bidder'.
(c) Responsibility for training the operating, servicing, and maintenance personnel rests largely with the customer.

(d) Improvements in design to improve reliability as a result of service experience can only be introduced by going through a complicated and slow 'modification system', which is usually under the sole control of the customer.

(e) As an important consequence of the foregoing points, it is not possible for the customer to hold the contractor entirely responsible for troubles arising in service.

Two items deserve comment:

1. COMPETITIVE TENDERING This practice is widespread throughout capitalist countries, since the Government agency must be seen to be making the effort to obtain maximum value for the taxpayer's money. Equally widespread are the criticisms voiced by contractors who find themselves underbid by competitors whose products they know to be inferior to their own. The users are often no less critical.

There is the oft-repeated remark made by John Glenn, the first American astronaut, at the press conference which was held after he had completed his mission. A reporter asked him what his thoughts and feelings had been during his $4\frac{1}{2}$ hour orbital journey. He replied, 'How would you feel strapped into a machine which has thousands of parts each supplied by the lowest bidder?'

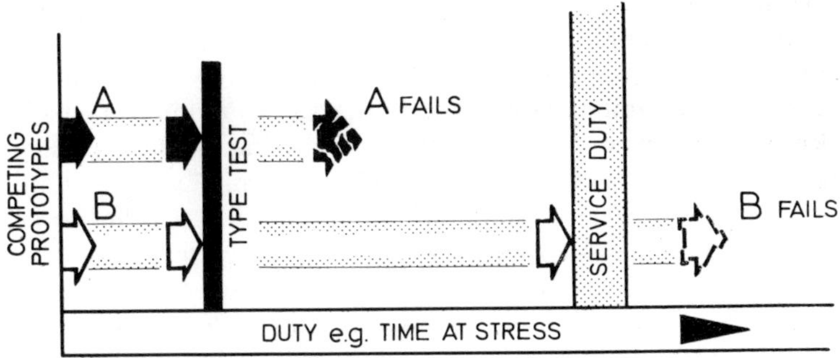

COMPETING PRODUCTS PASSING THE TYPE TEST RATE EQUALLY
BUT WHAT ARE THEIR REAL CAPABILITIES ?
AS A OR AS B ?
AND WHAT ARE THE CAPABILITIES OF ALL PRODUCTION UNITS ?

Fig. 5.1. The Fallacy of the Type Test.

Quite obviously there would be full justification for the placing of orders with the lowest bidder if the competing products were equally good. Lacking evidence to the contrary, the procurement branch of the

Government agency is not to be blamed. The party responsible is the technical branch which draws up the qualifying test.

All too often this consists of no more than a type approval test, of a specified duration, under specified conditions, usually on the test bed or in the laboratory. For reasons which will be discussed more fully in chapter 8, such a test can do little to demonstrate the potential reliability of the product. In the meantime, Fig. 5.1 will give sufficient indication of the basic argument.

2. GUARANTEED PERFORMANCE Some Government agencies in Britain and in the United States are making the attempt to buy better value, based upon the 'total life cost' of the product. The Navy Department of the British Ministry of Defence is offering Incentive Contracts for performance better than specification.[6] The American Department of Defense is actively studying methods of estimating the total lifetime cost of defence equipment—its first cost, its replacement and maintenance and failure costs, and so on.

Both of these approaches require closer collaboration between customer and supplier. They lead to the possibility that the supplier will be able to accept a greater share of the responsibility for satisfactory performance.

Industry-to-Industry Business Where the customer is an industrial organization, the operator of a fleet of aircraft or hire cars or buses, of machine tools, or of processing plant, a different attitude prevails.

The customer is immediately and directly conscious of costs. He usually knows from past experience how serious can be the consequences of breakdown in the equipment upon which his business depends. Increasingly, he is entering into a dialogue with his supplier to ensure that both parties benefit from a joint study of the conditions required to be met, and the problems which arise from these.

Where equipment is of a new type, the customer should demand proof of its reliability. He should involve his own technical staff, and he should negotiate with his supplier the additional costs of proving and demonstrating the effectiveness of the new design. These costs will be only a small fraction of those which could be incurred through breakdown of the equipment, and the consequent delays in achieving productive and profitable use of the plant or the service.

In all cases the customer is entitled to expect of the supplier that he will back up his product by adequate support. In particular, he should be able to expect immediate availability of spare parts. The example set by Boulton & Watt in the seventeen 'nineties should be followed.

At the same time, the customer must realize his own responsibilities.

When trouble does arise, he should be prepared to cooperate fully with his supplier in effecting a remedy.

When the customer is a military agency, considerable delays are usual in the acceptance of design changes. This is due in part to the complex nature of the agency's organization, to the lack of defined responsibility.

A civil airline will estimate the cost of living with a trouble, due to delays, frustrated passengers, loss of revenue, maintenance, etc. It will be quick to accept the cost of a modification which will eliminate the trouble.[7]

Other industries which may not be so aware of the costs incurred would be well-advised to compute them. They may be as high as those described by Sir Ronald Holroyd.[8]

For there to be the fullest cooperation between a customer and his supplier makes sense. Yet seldom do the relevant departments of the two contracting parties meet. Salesman meets buyer, but does engineer meet engineer?

Figure 5.2 reproduces slides which were used at a joint meeting between the Aero Engine Division of Rolls-Royce Ltd, and a gathering of its machine tool suppliers.

Customer responsibilities are no more than recognition of the fact that satisfactory performance results from an effective total system. This comprises not only all those involved in the supply of the equipment, but those involved in the specification, procurement, and use of the equipment as well.

The Supplier's Responsibilities

For his part, the supplier must do his best to evolve a product which will satisfactorily meet his customer's needs. The basic elements have been laid down on page 35. It will help if these are reproduced as definite activities.

The company undertaking to manufacture components or products designed by its customer must:

(a) Check that it understands what is required to be done to meet the requirements of the design.
(b) Assure itself that it possesses the skills and equipment which will enable it to meet those requirements.
(c) Confirm that the standards to be achieved are unequivocally defined in the customer's specification.

QUALITY AND RELIABILITY OF MACHINE TOOLS

WHAT FAULTS DID WE FIND ?

- FAULTY DESIGN
- POOR QUALITY MANUFACTURE
- DELAYED SERVICE
- POOR SPARES AVAILABILITY
- INADEQUATE DOCUMENTATION
- LATE DELIVERY

WHAT CAN YOU DO ?

★ IMPROVE AND DEVELOP DESIGNS

★ IMPROVE AND CONTROL MANUFACTURING QUALITY

★ IMPROVE EFFICIENT BACK UP SERVICE FOR ALL PRODUCTS

WHAT ARE ROLLS-ROYCE DOING ABOUT IT ?

WE ARE TAKING STEPS LEADING TO :

★ MORE PRECISE DEFINITION OF OUR REQUIREMENTS

★ MORE SYSTEMATIC AND RIGOROUS TESTING OF MACHINES BEFORE PURCHASE

★ INSTALLATION OF AN INFORMATION FEEDBACK SYSTEM TO MONITOR AND REPORT ON MACHINE TOOL PERFORMANCE

WHAT CAN WE DO BETWEEN US ?

★ ESTABLISH MUTUAL COOPERATION LEADING TO IMPROVED EQUIPMENT

Fig. 5.2. Customer/Supplier Cooperation.

(d) Reach agreement with the customer on all matters affecting the product, its price, its delivery, the supplier's responsibilities.

(e) Require that all this information is contained in a mutually agreed and formal contract.

(f) Procure materials, products, and services from its own suppliers of quality required to satisfy the requirements of the design.

(g) Plan methods of manufacture and control to ensure the manufacture of goods of the right quality.

(h) Manufacture components and products in accordance with the specifications and conditions laid down in the contract.

(i) When required, provide evidence to the customer that the necessary quality has been achieved, so that he can accept the product with the minimum of additional inspection.

(j) Deliver the products at the time promised.

(k) Accept responsibility for the quality of the goods delivered.

The Supplier as Customer Management which has been subjected to the tonic exercises laid down by, for example, the Navy Department's CP form 161 (General Requirements for the Assurance of Quality in New Construction Surface Warships) or its equivalent, will not unnaturally begin to wonder to what extent it can pass some of the burden on to its own suppliers.

The latest specifications, to which reference has been made, require a prime contractor to carry responsibility for all components and accessory products obtained from his sub-contractors and vendors. It is incumbent upon the contractor to ensure that his suppliers are as capable of satisfying him as he is expected to satisfy his own customer.

Parenthetically, there is the exception—the case where a Government agency procures subsidiary equipment directly from the manufacturers, and makes it available to the main contractor under a system going under the name of embodiment loan, or agency supplied items. Since this practice tends to absolve the contractor's management from responsibility, it need not be considered here, save to comment that the system suffers because of the lack of defined responsibility, or at best, of split responsibility.

When we consider commercial business, it is not usual for the customer to expect more of his supplier than that the equipment as a whole will be satisfactory. Surprisingly and regrettably, many managements fail to see that they are responsible to their customer for the quality of bought-out materials and parts. They are apt to disclaim responsibility for troubles experienced by their customer due to accessories not of their own manufacture. Some, indeed, have been able to divert the responsibility directly to their supplier, disregarding the fact that the way in which they themselves have installed the accessory, and the duty which they have imposed upon it, are directly their concern.

Enlightened management treats its suppliers in much the same way that it is treated by its customers.

In general, however, industrial management usually has a somewhat wider choice of supplier than is available to British Government agencies. Faced with a choice, it is important that management should make certain that its purchasing department is familiar with the techniques of vendor assessment. These are simple enough to apply, requiring as a rule no more than the application of a check list; a survey of the facilities of the vendor; an assessment of his apparent integrity, and a measure of his performance, based upon the quality of goods received. This aspect of management is dealt with more fully in chapter 13.

The Contract

If the relationships between supplier and customer, customer and supplier, are to be conducted satisfactorily, they must be established on a formal basis, by means of a contract. For far too long, certainly for a period of time which coincides with that during which the inspection function has been misunderstood and debased, the contract itself has been regarded as a mere formality. Excess of demand over supply, a situation long prevalent in Britain, has given the supplier the edge over his customer. Few if any sanctions have been applied when the supplier has failed to conform, as often as not because the contract itself had failed to specify what was of basic importance—the quality of the product to be delivered.

Much has been written on the subject of purchasing contracts. Aspects such as quantities and rates of delivery, specifications, price, terms of business, time of delivery, special conditions, inspection, guarantees, and penalties are dealt with in some detail. Usually, however, and in line with the general trend, little is said about the quality of conformance. The system necessary for the achievement of the required quality, the evidence that the quality has in fact been achieved, the conditions of acceptance inspection, the conditions for dealing with unacceptable products, the methods of submission of non-conforming products, and so on—these will only be found in the quality assurance requirements of some few large organizations, and in the purchasing contracts of even fewer.

Managing Affairs Efficiently

It may have seemed from the foregoing that top management is being saddled with a host of additional responsibilities. On the contrary, the

70

customer- and product-oriented approach to the overall purpose of the enterprise leads to such a clarification of vital issues that the overall efficiency of the organization is increased, and management's burden is lightened.

The first step is the important one: to ensure that all the departments in the organization know what their own responsibilities are, that they have the ability and the facilities needed to meet those responsibilities, and that they carry them out properly. In all probability there will exist already a sound organizational pattern, which will, however, lack the essential cross-links which the system demanded by the quality assurance requirements of various large customers provides. Industrial management has cause to be grateful to the US Department of Defense, to the British Ministry of Defence (Navy Department) and to NATO's Military Agency for Standardization for having laid down so clearly the essential steps towards the achievement of requisite quality of product.

Not the least valuable contribution has been the implied necessity for clear specifications, at every stage from the definition of need, through every phase of design and manufacture to the finished product.

References

1. 'Quality Program Provisions for Aeronautical and Space System Contractors.' NHB 5300 4(1B) April 1969. National Aeronautics and Space Administration.
2. Roberts, Thomas S. 'Ideal technical–managerial mix exists for all supervisor levels.' *SAE Journal*, Vol. 77, No. 11, November 1969.
3. Derived from *The Golden Book of Management*, ed. L. Urwick, Newman Neame Ltd., 1956 London. See also Roll, Eric *An Early Experiment in Industrial Organization*. Frank Cass & Co. Ltd., 1930 & 1968, the original source.
4. George, Claude S. Jr. *The History of Management Thought*. Prentice-Hall Inc., Englewood Cliffs, N.J., 1968.
5. AQAP-1 and AQAP-2 are published in Britain by the Ministry of Defence as MoD Specification 05–8.
6. Hedger, *loc. cit.*
7. Bowling, A. G. 'Lessons from engine experience.' Colloquium on Aircraft Reliability in Service. *J. Royal Aero. Soc.*, Vol. 70, No. 663, London, March 1966.
8. Holroyd, Ronald 'Ultra large single stream chemical plants: their advantages and disadvantages.' *Chemistry and Industry*, London, August 5, 1967.

6
The Specification

The *Oxford English Dictionary* defines 'specification' as 'A detailed description of the particulars of some projected work in building, engineering, or the like, giving the dimensions, materials, quantities, etc. of the work, together with directions to be followed by the builder or constructor'.

There is perhaps significance in the fact that this definition dates from 1833, the time when, with the beginning of the railway age, engineering was establishing itself as the basis of modern industry.

The need for specification of requirements extends into every sphere of activity. At the one extreme, there may be the description of the need for a new item of defence equipment which may lead to orders worth hundreds of millions of pounds. At the other, there is the simple verbal instruction given by a chargehand to an operative, telling him what he is required to do. The specification may be a written or a verbal communication, but more and more it is coming to be appreciated that the risk of misunderstanding is lessened if needs or instructions are set down in writing. To produce a clear and unequivocal written instruction of requirements is not an easy task. Some people have gone so far as to query the feasibility of producing specifications which will be absolutely clear and definitive in black-and-white terms. Others feel that in consequence an incomplete specification is not worth the effort, being worse than no specification at all. This, however, is tantamount to saying that it is better to remain in ignorance than to try to throw at least a little light on to a dark situation. The cost of the effort will be more than recovered by avoiding the losses due to misunderstanding and incomplete instructions which are otherwise inevitable. Moreover, as

72

will be shown, simple guide lines can be laid down which greatly facilitate the preparation of meaningful specifications.

Specification, the written or spoken description of requirement or need, is the basis of effective communication between two parties. Accurate and effective communication becomes increasingly important as products become more complex and as our dependence upon their correct functioning increases; as industrial organizations grow in size, and depend upon increasing numbers of other firms for materials, components, and services. In consequence, it is incumbent upon management, today as never before, to ensure that there is someone in the organization responsible for the preparation of specifications, and that he is given the opportunity to practice and improve his art.

It will be recalled that in 1935 Shewhart laid down as the first requisite for the control of quality 'the specification of the aimed-at standard of quality'.[1]

If further justification of the importance of the art of specifying were needed, one could adduce again the example of Marks and Spencer Ltd. This company's highly successful policy of quality control has resulted in a continuous raising of value to the customer, and sales and profits have increased rapidly since the policy was adopted. The basis of this policy is clearly written specifications of the products which are required from the company's suppliers. After confirmation that the supplier has understood the specification of a product, and has demonstrated his ability to meet it, he is then made responsible for the quality of all future deliveries of that product. His performance is monitored by occasional checks, but more especially by careful counting of customer returns.[2]

> Some years ago the late Lord Marks was describing to the present author how his company had come to realize the importance of the specification. At the time he was concerned with the procurement of garments from sub-contractors, and he told how the specification for a shirt included such details as needle size, count of thread, number of stitches per inch. Straightness of lines of stitching, consistent width of seams, the quality of the stitching-on of buttons were stipulated by samples produced by young girls with no more than three weeks' training. This ensured that the supplier was being asked to do no more than was within his capability.
>
> Lord Marks asked if the specifications of sub-contracted machined parts included similar detail. Did they, for instance, include instructions regarding tool design, speeds, feeds, and coolant. He

was told that the engineering industry differed from the 'rag trade' in that sub-contractors were expected to know these things, so that it was not considered necessary to specify them. Lord Marks' reply was 'You will have to one day'.

When we stop to consider the costs which are incurred by the failure of sub-contractors to deliver goods to the required but usually unspecified standards of quality, at the rates and on the dates required, it can only be concluded that Lord Marks showed a prescience of a situation which must inevitably come to pass, and the sooner the better.

The Purpose of the Specification

The overriding purpose of a specification is to ensure that, as far as is humanly possible, the two parties to a business deal know:

1. What the party of the first part requires of the party of the second part.
2. What the party of the second part believes will satisfy the party of the first part.

Mutual satisfaction requires that there shall be a contract between the two parties, customer and supplier, which must be based upon prior understanding and agreement on the following points:

(a) The general *terms of business*.
(b) The *product* or *service* which is required, with stipulated standards of quality, or of performance.
(c) The dates and rates of *delivery*.
(d) The *price*.

In unpublished talks E. J. Goodall, Chief Buyer of the Aero Engine Division of Rolls-Royce Ltd, has pointed out that contracts have tended to become one-sided. The interests of the supplier must be borne in mind, and the contract should take into account such factors as risk, opportunity, penalty for breach, and reward for performance.

These are important aspects of the partnership between supplier and customer which it should be the aim of every buyer to encourage. A contract which is based upon mutual understanding of the interests of both parties is essential to such a partnership.

Future misunderstandings of some or all of the customer's needs will be avoided only when each of the foregoing items is specified so fully and clearly that there can be no reasonable risk of misinterpretation; when the consequences of failure to conform with the conditions are laid down; and when agreement is reached on price changes which might be necessary to match changes in costs which are outside the control of either party.

Agreement having been reached upon all the factors involved, the supplier then becomes entirely responsible for meeting his part of the bargain by satisfying his customer on all counts.

Before considering the contract and the product specification in detail, it will be useful to discuss the various kinds of specification.

Kinds of Specification

The purpose of a specification has been seen to be to establish a clear understanding between two parties. It is unlikely that any two parties will have, initially, an identity of viewpoint. While not necessarily in opposition, indeed one party will usually want to buy or receive goods or services and the other to sell or deliver, there will usually be the need to reconcile some differences. For example, a performance may be required of a prime mover, or of a machine, which is not at the moment attainable. Should the customer not declare his desires, should the supplier not state what he can achieve, and should no compromise performance be agreed, the result could only be disappointment to both parties. The attempt to lay down a specification of requirements and to have it accepted by both parties is an excellent way of bringing to light any differences which might otherwise remain unrecognized until trouble arises.

Although given different names, specifications are required at every stage in the evolution of a product from the initial concept of a need, to its correct use and maintenance in service. They will vary widely in type and form of presentation, but the same general principles apply to all of them. Specifications may exist in the following forms.

1. THE INDICATION OF A NEED This may be no more than a board meeting minute suggesting that a market exists for a certain type of product, and asking that an investigation be made; or it may be an invitation issued by a Government department, asking industrial firms to prepare design studies for a new type of product. It will usually be expressed in general terms, but where possible the performance, the conditions of use, and the environment should be indicated. This is in line with the customer responsibilities which were discussed on page 62.

2. A DESIGN PROPOSAL This will usually be no more than an outline study. It should define the basic design, the intended performance, the probable cost, and the estimated completion date.

Alternative proposals might be submitted, and depending upon the degree of technical innovation, a certain amount of research and development might be required, to optimize the factors influencing performance and to confirm the feasibility of the schemes. This research and development might be carried out by the company, or if the project were entirely novel, a contract to carry out preliminary investigations might be awarded by the prospective customer.

3. COMPLETE PRODUCT DESIGN When the final form of the product has been decided, a complete design can be prepared. This in itself is a specification, which describes, usually by drawings, the arrangement of the product, the size and shape of detail components, the materials to be used.

4. PRODUCTION DESIGN The final schematic design needs to be converted into detailed instructions to the manufacturer. These instructions usually comprise a complete set of detail drawings of every component, fully dimensioned, with details of material and finish. Today, all but a minority of engineering components have toleranced dimensions. These tolerances are themselves a specification—of the variation in size which will be acceptable.

5. PRODUCTION INSTRUCTIONS The specification of the design which is conveyed to the manufacturing department in the form of detailed descriptions of components must be translated so that it will have meaning to those whose function is to convert drawings into hardware. Forging drawings, pattern and casting drawings, drawings of components at intermediate stages of operations, process instructions, inspection instructions, the design and calling up of appropriate jigs, fixtures, and tools comprise the manufacturing specification. It is an essential part of good housekeeping. As a rule its preparation is carried out better than is done in some of the other areas mentioned. This is probably because the production engineer and the planning engineer are already working closely with the designer on the one hand, and the producer on the other. Faced with the responsibility for delivering the goods, requirements can be seen more clearly.

6. QUALITY STANDARDS For complete definition of a component, the minimum level of quality which is acceptable must be laid down.

In the case of a simple machined component, this is defined by the tolerances on dimensions. Here the detail drawing is the quality standard. It indicates the maximum deviation from the norm which is acceptable.

The quality standard should, as a matter of principle, demand no higher level of quality than is necessary for the correct functioning of the part. Unfortunately, often through ignorance of the capabilities of the machining processes, designers tend to specify much tighter tolerances than are necessary. At best, this results in higher costs of manufacture. At worst, it leads to friction between manufacturing, inspection, and engineering departments, and lowers the prestige of the specification, which comes to be ignored. This situation is common in every industrial country.

In the case of non-dimensional attributes (for example the porosity of a casting, the non-metallic inclusions in a critically-stressed forging, the porosity and appearance of a weld bead), it is not usually possible to lay down standards at the time the design is prepared. This is because no one can really know before production begins the extent of the variability in process capability which is likely to be experienced with the components in question. Separate quality standards must be laid down, and this is a responsibility of quality engineering (see chapter 14, page 191).

In the cases of subjective or qualitative standards, it may be necessary to specify them by means of samples, models, or photographs. An impressive example is the uniformity of fruit and vegetables—apples and lettuces come to mind—which are offered for sale at the better super-markets. Wherever possible, however, standards of acceptance should be defined quantitatively, in numerical terms.

7. THE PRODUCT SPECIFICATION When a product is manufactured entirely in a company's own plant, and sold to customers of its own seeking, the specifications outlined above will be sufficient to meet the needs for communication between departments.

In engineering manufacture it is found, however, that an average of 60 per cent of the finished product, by value, is obtained from outside sources, in the form of raw materials, finished components, or pro-prietary accessories. There is too the frequent case where a firm is manufacturing to meet the requirements of a single large customer. In these cases a complete product specification is necessary.

It will comprise:

A description of the product.
Its purpose.
Its performance.
The design.
The detail drawings.
Assembly drawings.
Material specifications.
Process specifications.
Specifications of manufacturing techniques.
Quality standards.
Inspection methods.
Acceptance tests.

The product specification along these lines is an essential part of the purchase contract.

At a different level, as when the customer is the domestic consumer or the motorist, the product specification usually takes the form of a glossy brochure, or an attractively illustrated hand-out. Even though its purpose may quite overtly be to attract sales, it is essential that it must be truthful and accurate. The customer is now protected by the Trade Descriptions Act 1968.

8. THE USER SPECIFICATION Satisfactory performance of product depends upon its being used properly. To ensure this, the wise supplier will provide clear instructions so that the customer will get best value from his purchase.

A simple everyday example is the method of application which is printed on the bottom of a shoe-polish tin. With medicines, the instruction may need to include a warning regarding the dangers of an overdose, and this will usually be on the label on the bottle. In the case of a radio or a television set or other item of domestic equipment, the specification of the way in which the product should be used may be no more than a small leaflet simply indicating the position of the controls. At its best it will be a small booklet, well-illustrated, which includes advice on care and maintenance, and warnings against repair of electrical components by unskilled people.

This wide variation suggests that not all manufacturers have realized the importance of the user as a factor in achieving satisfactory performance of the product. The same tendency is evident in respect of motor cars. The manual with which the owner is provided is often today no more than the briefest description of the controls, the types of sparking plugs and lamp bulbs, the approved oils and frequency of oil change. It is not unlikely that the majority of motorists would welcome instruction and advice on tyre care, on how to drive to achieve best fuel economy, and longest car life, how to preserve the bodywork, how to tune the carburettor, and adjust the door locks.

On the other hand, most washable garments today carry an informative label giving advice on the preferred method of laundering and ironing. This became a necessary part of user education as more and more new fabrics were put on the market.

Manufacturers of more costly and complex equipment treat this matter seriously. Whole departments exist to compile instruction manuals which for a ship's machinery, an aircraft, a processing plant may run to several large volumes. Managements of concerns involved with less costly equipment would nevertheless be wise to follow this example, adjusting the effort of course to the type of product and to the importance of the market.

Specifications in this area of activity present a special challenge. The reader/user is seldom able to consult the author on matters of interpretation, and in many cases he is not an expert buyer or seller, designer or producer. He is merely the most important man in the business cycle, the customer.

In the better examples of instruction manuals, effective use is made of diagrams, sketches, and photographs. It goes without saying that the written matter should be in plain simple everyday language, with none of the jargon to which the producer will be accustomed, but of which the user cannot be expected to be aware.

The Strength and Weakness of Standards

Writers on specifications tend to suggest that standards should be incorporated wherever possible. Unfortunately, the words 'specification' and 'standard' are used loosely, often interchangeably. Here we are discussing standards which have been promulgated by a national body (e.g. the British Standards Institution) or a trade association or professional institution (e.g. the Society of Automobile Engineers).

It is true that many products can be standardized, with great overall economies. The wider adoption of such standards can do nothing but good. We are inclined to take for granted the convenience of having standard electric light fittings, voltages, and wattages; standard power plugs, wiring, switches, and conduits. Users of electric razors who travel abroad know of the difficulties due to different standards of power plugs and appreciate the value of standards all the more when they return home.

Buyers of materials for industry greatly value the convenience of standard sections, gauges, and materials. As long as designs remain fairly static, so that new requirements are not posed, standardization is a very good thing. However, if technical progress is not to be hampered, it is essential that standards should be kept up-to-date by continual review.

It is when we come to areas of industry in which rapid technological advances are being made that standards become frustrating, mainly by reason of the inordinate time which appears to be required to evolve new standards, and to update old ones. These aspects have come under serious critical review by the Institute of Purchasing and Supply.[3] Criticisms are levelled as well at Government departments and nationalized industries, for writing their own individual specifications for what are similar materials and items of equipment. This, however, may to some extent be excused by the lack of currently available specifications suitable for their needs.

It is when we come to specifications of materials of construction, however, that there are the strongest grounds for complaint. The present author has referred in numerous papers to the need to apply the same criteria to metal specifications as are applied to finished products. Indeed, as Dr Kinzel[4] has pointed out in his lecture great advances in design could be made, with enormous economies in materials, if specifications called for quantified standards of quality, declared values of the range in physical properties, and data on fatigue properties.

It should be possible to use a standard as the basis of a purchase agreement. The vast majority of metal specifications fail to meet the first essential requirement—of precision. To describe qualities of bar by such terms as 'to be free from laps, seams and roaks', of castings 'to be free from porosity', and so on is merely to sow the seeds of discord between supplier and customer. How free? Detected by what method? It should occasion no surprise that manufacturers of products of integrity find it necessary to lay down their own requirements.

What is more astonishing is that suppliers of metals should make such a small effort to understand and to meet the needs of their customers. There seems to be completely lacking, in all countries, that sense of responsibility towards the customer which is possessed by all suppliers of satisfactory products. Exceptions known to the author are two American steel forgers. It is perhaps significant that neither has deep roots in the business, and has therefore not developed the traditional outlook. One started as a brewer, the other as a builder of oil-well drilling equipment. Both were compelled to produce forgings to meet their own requirements because the requisite quality was unobtainable from other sources. They brought a fresh approach to the task. Both became expert, and pioneered new techniques of manufacture of forgings which have had important applications.

Having been customers themselves, they make the requirements and the satisfaction of their customers their especial concern.

It is only a little less surprising that manufacturing industry has for so long been content with this situation. The cost of maintaining checks of the quality of incoming materials, by all manufacturers who cannot afford the consequences of defective material, must be truly enormous.

This point was given rare emphasis by Dr Augustus B. Kinzel in a strongly worded address to the American Society for Testing and Materials:

> Specifications?!—The question mark pertains to the first part of this lecture—What is the purpose of a specification? . . . The

exclamation point pertains to the astonishingly unsatisfactory
state of the art, to the astounding lack of communication and
pertinence in many specifications, and to the inexcusable waste
due to ignorance of materials, meaning of tests, and lack of
engineering service requirements. . . .[4]

Even more important is the need to include data on the fatigue
properties which can be guaranteed, so that the material supplier can
assume his due share of the responsibility for reliable end-product.

It will be seen when we discuss the relevance of fatigue properties to
the achievement of reliability that a knowledge of variability, of the
scatter in fatigue strength and fatigue life, and in other physical properties, should be available. This is necessary to enable the designer to know
what is the worst that he may expect in production quantities of material.
It is surprising that with rare exceptions material specifications contain
little or no indication of the standard deviation in the numerical values
of physical properties.

> Yet as long ago as 1935 Professor Egon S. Pearson prepared for the
> British Standards Institution a valuable specification on the use of
> statistics as a measure of the variability of physical properties of
> materials.[5] The unfortunate accident of the loss of the set type in a
> bombing incident during the war is no excuse for the continued
> failure to make any attempt to include in specifications the mean
> (x) and range (R) of measurable properties.
>
> This matter received further notice in 1938 when W. T. O'Dea
> pointed out the fallacious nature of specifications based upon
> average values. Figure 6.1[6] shows how a batch of product having
> a much lower average value can still have a higher minimum
> value than a batch apparently much better on the basis of its
> average value only.
>
> Yet the designer needs to know the *minimum* value of the
> properties of the materials which is likely to be encountered in
> production batches. Unless he can base his calculations on this
> knowledge, it is highly likely that some failures will occur in
> service. Lacking this information, the designer is compelled to
> apply 'factors of safety', which are more correctly described as
> 'factors of ignorance'.[7] [8] In consequence, the design lacks refine-
> ment and is wasteful of material.

It would appear, therefore, that for as far ahead as can be foreseen,
the use of national standards should be restricted to those applications
where existing static designs are acceptable. In all other cases, standards

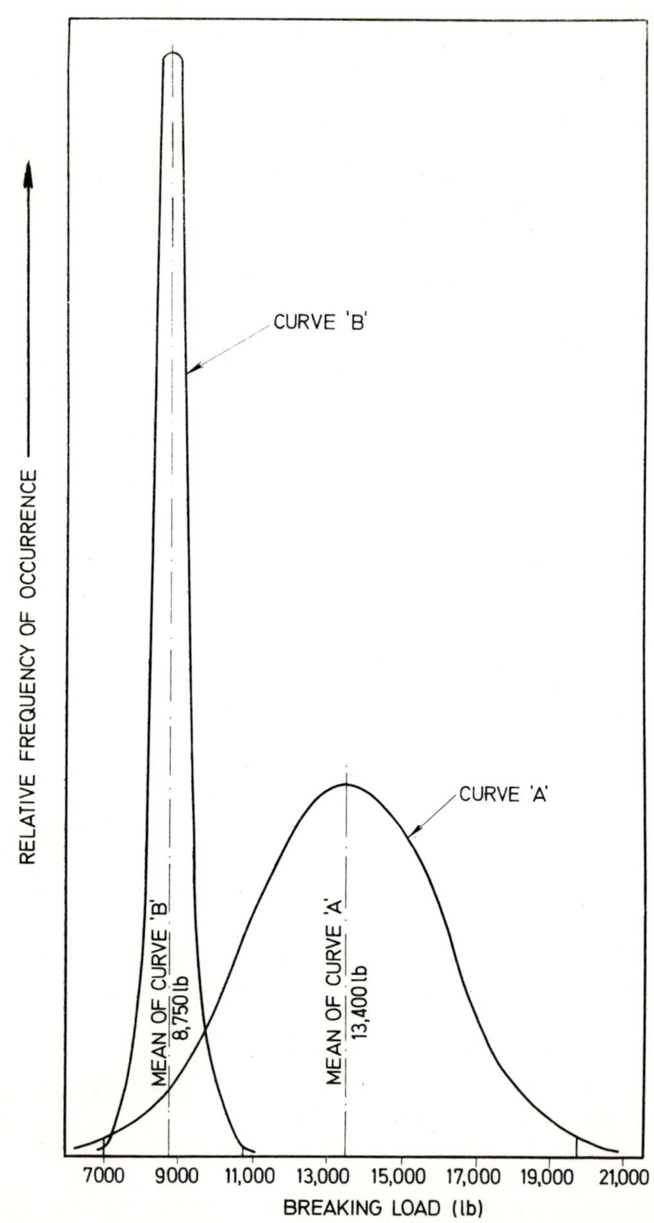

Fig. 6.1. For a Specification to be Effective, *Mean* and *Range* must be Quoted.[6]

can only lead to stagnation. With materials of construction, it is to be hoped that a better understanding of the fundamentals of specification writing, and a wider appreciation of the responsibilities implied in purchase agreements will lead before long to specifications which, by quantifying quality stipulations, can become the basis of legally enforceable contracts.

The Specification of a Specification

A specification is a document intended to convey from one person or party to another a clear understanding of what the one requires of the other. The requirements may be a product or an activity. The activity may be that of designing a product, or of machining components to the designs of the customer (sub-contracting), or it may mean the provision of a service. Ideally, it should define the minimum which needs to be done to satisfy the needs of the customer. This should be written in such a way that the specification alone is sufficient to convey to any prospective supplier a clear idea of what is needed to satisfy the requirements of the design. Only by this means can ambiguity be avoided. Another matter, of great importance to firms whose products have achieved an initial success, and to Defence departments, is the assurance that the original product can be reproduced exactly, at any future date, by any manufacturer.

A specification is, therefore, a precise definition, or as precise a definition as can be written at the time. It is a means of reconciling the views of the two parties to an agreement, since on the one hand it must define the conditions which will satisfy the customer, and on the other it must be accepted by the supplier as being capable of achievement.

There follows a list of items which should be included in a fully comprehensive product specification. Many products can be specified by something less than this. Minor components, for example, can be completely specified by the detail drawing.

(a) The *form* of the product.
(b) The *purpose* of the product.
(c) The *description* of the product, including the material from which it is to be made, its dimensions, tolerances, finish.
(d) The *method* of making the product. This will usually be defined in subsidiary process specifications, manufacturing technique specifications, operation planning sheets.
(e) The *quality* of the product, i.e. the minimum level of quality which will satisfy the requirements, e.g. of porosity, cleanliness. This will normally be defined in a quality specification.

(f) The *inspection* of the product. Again, this will usually be covered by a separate specification or specifications, where special methods, e.g. of non-destructive testing, or pressure testing, are required. The checks to be carried out by supplier and customer must be laid down.

(g) The *performance* of the product. This will apply particularly to functioning devices, e.g. engines, fuel systems, actuators, etc. Minimum acceptable performance, and methods of testing, must be specified.

(h) The *reliability* of the product. This will stipulate the degree of reliability under defined conditions, with the method of estimating or measuring this.

(i) The *durability* of the product. This will cover the total useful life expected, with the means of measuring it.

(j) The *packaging* and *storage* of the product. Description of method of packaging and any special requirements of storage.

In 1965, the Swedish Association of Metalworking Industries (Sveriges Mekanförbund) published, in English, a memorandum 'Preparation of Technical Specifications for Use in Purchase'. It was the first contribution to the subject which had appeared for many years. It was followed in 1967 by a 'Guide to the Preparation of Specifications'.[9] This last was prepared by an *ad hoc* Committee of the National Council for Quality and Reliability, and published by the British Standards Institution. Both documents are useful as reminders of what should be included, although, regrettably, neither sufficiently emphasizes the importance of quantifying all measurable attributes.

Writing the Specification Since the aim is mutual understanding, clarity and simplicity are prime requirements. Clarity of definition needs plain, simple language and the *specification* should be:

(a) As short as possible. Unnecessary words or descriptions lead to lack of precision.

(b) Positive in meaning and intent.

(c) Unambiguous.

(d) Specific and precise.

(e) Free from vague generalities such as 'to be of good finish' or 'to be free from blemishes'. No question such as 'how good?' or 'how free?' should remain unanswered.

(f) Numerical wherever possible, e.g. finish should be specified by the CLA (centre line average) value.

(g) Clear regarding the *minimum* acceptable level of quality or performance. (*Note:* This may be effected by specifying the *average* and the *range* of characteristic required, as with a nominal dimension and the tolerance. In the case of non-dimensional attributes, e.g. porosity, magnetic particle inspection indications of non-metallic inclusions, etc., the minimum acceptable level will be defined by the maximum size and number of blowholes, and so on, in specified areas.)

(h) Augumented by photographs, diagrams, models, or samples where a precise numerical value cannot be given.

(i) Free from the common fallacy of calling for an unattainable 'desirable' target.

(j) Supported whenever possible by national or company standards, but only after it has been confirmed that the standards themselves are clearly expressed and precise in their meaning.

During a talk which he gave in London in March 1969, John J. Riordan, Director of Technical Data, Standardization Policy and Quality Assurance in the US Department of Defense, said

In all the industrial countries of the world, if there is any area of technology which can be considered an underprivileged or under-developed area, it is the field of what I would call specification technology. This is a fact of life which we all have to live with and this incidentally is verified by the fact that when one goes to the Library of Congress to get a book on specification writing there is no book. There are fragments of books, but a measure of the cultural development of a subject is its literature, and there is very little really good literature on this business of how to write a specification. I think that it is possible to write definitive specifications for such items as clothing or common stores. I also think it is possible to write definitive specifications for items that are in a process of change, for example, capacitors, resistors and so forth. But as one gets into these more changeable and evolving items, the requirements are increasingly in performance terms rather than in descriptive or dimensional terms. My answer is yes, it is possible, but it takes a real talent and most Governments and most industries do not put their best people into the business of specification writing. The point is that the problem of specification is central to business management but as I see it neither in Government nor in industry (I'm speaking of course of American Government and industry) has there been recognition that the man who writes the specification has a blank cheque on the Treasury. We do not realize this, but the fellow who puts a particular test or a particular requirement in a specification could be committing the U.S. Government to vast sums of money. But for some reason this is not understood and in consequence there is no real major effort to support the preparation of good specification.

It is obvious from the foregoing that an essential qualification of the specification writer is that he should be well-qualified and experienced in his subject. He must be able to visualize how the specification will be interpreted by those who must implement it, and he must have a sufficiently broadly-based experience to avoid the pitfalls of excessive requirements of proof of quality mentioned by Riordan.

A useful guide to the writer of specifications was compiled over 40 years ago by N. F. Harriman.[10]

Precepts for Specification Writers

(a) Either *specify* or *omit*. Eliminate all clauses that reflect indecision or lack of knowledge. Do not put anything in the specifications that will not be enforced to the letter.

(b) Use simple words. Use technical words in their *exact* technical meaning. Do not use words that are subject to ambiguous interpretation.

(c) Use nouns. Do not use pronouns. It is better to repeat the nouns, even at the sacrifice of elegance.

(d) Do not write long and involved sentences.

(e) Use commas sparingly. Do not construct sentences in such form that the omission, addition, or misplacement of a comma will alter the sense.

(f) Make the language a clear and concise expression of just what is meant.

(g) Give directions, not suggestions. Tell the manufacturer *definitely* what shall or shall not be done.

(h) Do not attempt to conceal obligations or to place all the risks on the manufacturer.

(i) Either specify *results desired* and leave methods and properties to the manufacturer, or specify *methods* and *properties* desired and assume the responsibility of results.

Finally, the specification writer will be well-advised to bear constantly in mind the rhyme of Rudyard Kipling:

> I keep six honest serving men,
> (They taught me all I knew);
> Their names are What and Why and When,
> And How and Where and Who.[11]

This advice is applicable not only to the writing of specifications, but to the solution of problems of all kinds.

The principles of specification writing which have been described could have applications of even greater value. Used in discussions between the two parties in industrial dispute, it should be possible to avoid much of the misunderstanding which is the root cause of so much industrial unrest.

References

1. Shewhart, W. A. 'Nature and origin of standards of quality.' *Bell System Technical Journal,* No. 1, Vol. XXXVII, January, 1958. Reproduction of a paper of 1935.
2. British Productivity Council film *Right First Time*. Obtainable on loan from the Central Film Library, London.
3. *The Engineer*, 24th April 1969, and *Purchasing Journal*, April 1968 (Journal of the Institute of Purchasing and Supply).
4. Kinzel, Augustus B., giving the Tenth Gillett Memorial Lecture, American Society for Testing and Materials, 1961.
5. Pearson, Egon S. 'The Application of Statistical Methods in Industrial Standardization and Quality Control.' BS 600, 1935.
6. O'Dea, W. T. 'Statistical methods and factors of safety: an argument for the revision of existing standards.' Institution of Electrical Engineers, Vol. 87, No. 523, July, 1940.
7. Nixon, F. 'Quality in engineering manufacture.' Institution of Mechanical Engineers, 1958.
8. Kinzel, Augustus B. *loc. cit.*
9. 'Guide to the Preparation of Specifications.' PD.6112, British Standards Institution, May, 1967.
10. Harriman, N. F. *Standards and Standardisation*. McGraw-Hill, 1928.
11. Kipling, Rudyard. 'The Elephant's Child' in *Just So Stories*, 1902.

PART 3

Evolving the Product Specification

7

The Engineering Function

A whole sequence of activities, coming under the general description of 'engineering' must be brought into action in order to convert a managerial decision to enter a particular market into a product which it is hoped will satisfy that market. These activities come under the headings of design, development engineering, production engineering, procurement, manufacture, and quality control. Here, we are concerned with the earlier stages, that is, those activities, of design and development, which are necessary to define a product which will satisfy the customer.

In many of today's more forward looking organizations, these two stages are combined under the general title of 'engineering'. This tends to de-emphasize the design activity, but it has the advantage of recognizing the essential partnership with development engineering. This is all to the good, since even where management recognizes the importance of good design there is plenty of evidence to show that the hardly less important role of development testing is not yet fully appreciated.

For reasons which are peculiarly British, although they exist to some extent in the United States, engineering has never been accorded the professional status which it receives in other European countries. To some extent, this may be due to the long persistence of the Greek tradition in our older universities, which regarded practical, technical, and scientific pursuits as ungentlemanly. This was strengthened by the exclusion of non-conformists from the universities of Oxford and Cambridge. Many of these men, escaping the stereotyping of an orthodox university training, turned to industry, where they played a great part in establishing Britain's lead in the 18th and 19th centuries.[1] Another reason is perhaps the word 'engineering' itself, which carried the connotation of engines and of the men who controlled the engines, men wearing overalls and greasy caps. It is a pity that to correct this misconception we have come increasingly to use the word 'technology'. This is in fact a less meaningful word than 'engineering', which stems from the Latin *ingenium,* the *ingeniator,* or engineer, being an ingenious man. This was recognized by the Institution of Civil Engineers 150 years ago, when they adopted Thomas Tredgold's definition

> Engineering is the art of utilizing the great sources of power in nature for the use and convenience of man. [2]

It will help management better to appreciate the importance of the role of the engineering department if this is borne in mind.

Current overuse of 'technology' and its too frequent combination with 'science' have led to a widespread impression that a technique exists for the solution of every problem. Just as there is art in engineering, there is art in management, and success in both engineering and management depends upon recognition of this fact. It is worth repeating again the theme of this book, that by focusing attention upon customer [3] and product, the Q & R concept leads inexorably to observance of these fundamental arts both of management and of engineering.

In this connection it has been stated that the Q & R concept has been concerned excessively with engineering products and the engineering industry. The fact is that 'engineering thinking' is fundamental to all human activity, and the ideas which are promulgated here are applicable to all branches of industry.

As we have seen, basic weakness of the design of a product carries by far the greater share of the responsibility for unreliable performance. In a period in which technology (used here in its proper sense of 'the scientific study of the practical or industrial arts') is advancing more rapidly than ever before and markets are becoming increasingly more difficult to win in the face of intensified competition, engineering, the 'art' content of the subject, assumes even greater importance. It is significant that the most successful manufacturing companies are those with strong and able teams of design and development engineers. Usually, the boards of directors of such companies include men with an engineering background, well able to appreciate the importance of the engineering function, and to weight it accordingly.

Since it is essential that adequate staff and facilities should be made available, an account will be given of the function of the engineering departments as a whole, and of the action which should be taken to ensure that the ensuing product shall have the best possible chance of

92

success. If at the same time this account gives help or encouragement to engineers themselves the effort will have been doubly justified.

The Cost of Engineering

First, let us justify the cost of engineering design and development. In the case of highly sophisticated products where each new model represents an advance beyond the frontiers of existing scientific and technical knowledge, the design and development work necessary to evolve a viable specification of product is admittedly expensive. This is, however, well understood by the managements concerned, and full account is taken of it in determining policy.

Taking as an example a new type of aircraft engine, the necessary research, design, and development can cost from 125 to 150 times the selling price of the engine which will eventually be produced. The provision of new manufacturing facilities, production planning, and tooling can cost as much again. To limit the amortization of these costs to 10 per cent or so of the price of the engine and of the spare parts which it will require during its useful life necessitates an assured market of at least 1600 engines.[3] This high cost of the launching of a project, and the limited market, are factors in the low profitability of some technologically advanced industries.

Less sophisticated products such as motor cars may command markets two hundred or more times as great, for a single model. In such cases, the design and development content of the cost of a single unit of product will be quite small, of the order of 1 per cent of the selling price. This leads one to wonder why there are such marked differences in the quality of the design of models of different makes within a given price bracket. To hire the best brains available would have a barely discernible effect upon the price of his car to a customer, yet the improvement in customer satisfaction, and the increased assurance of a greater share of the market would be out of all proportion.

Where the product is a single complex unit, such as a large scale chemical plant, a power station, a ship, or a bridge, the total cost may well amount to many millions of pounds. The cost of design and development, relative to the product, may loom large compared with the previous example. This may be a reason for the seeming neglect of the engineering phase seen in the too-frequent examples of catastrophically expensive failures of large one-off equipments during recent years. The consequential cost of the failure of a single component might easily run into millions of pounds, through penalties for delay, lost revenue, or diminished return on capital invested due to reduced performance. In addition, there is the incalculable cost of lost reputation and of lost prestige. Any conceivable increase in expenditure necessary to strengthen the design and development effort would represent a very small insurance premium indeed.

The Function of Engineering

Engineers, who increasingly are becoming specialists in narrow fields, tend to look upon their work as an end in itself. The manager must preserve the broad viewpoint. He should see that the function of the engineering department is to convert a corporate decision to cater for a particular market into clear and unequivocal descriptions and instructions. This will ensure the manufacture of products which will satisfy the needs of the customer, at minimum cost to the company. He must see the function of each engineer, of each section of the engineering department, as contributing to the evolution and the finalization of the product specification.

The total activity of engineering employs a wide range of abilities and talents. These include:

(a) Creativity, imagination, and the ability to visualize conditions of operation and of the form of products and components which will be able to cope with the conditions which might arise from the intended duty.

(b) The ability to synthesize a complex product from an assembly of detail components, and to analyse the interactions of these components.

(c) A knowledge of statics, dynamics, and kinematics.

(d) A knowledge of the properties of materials, of methods of manufacture, and of construction.

(e) An awareness of the economics of manufacture.

(f) Foresight of the conditions of use and abuse to which the product might be subjected at the hands of the user.

(g) The ability to describe what is required to be done by the manufacturing and quality departments in order to evolve products which will meet the needs of the design, through the medium of drawings, specifications, and standards of quality.

For the efficient deployment of this bewildering variety of talents there must obviously be a plan and a programme. Indeed, the management of an engineering activity probably presents a greater challenge to managers than they are likely to find in any other department of the organization. It calls for the management of men who are practising an art, much of which will be a mystery to a non-technical manager. As a result, all too seldom do we find engineering management recognized as an essential function in the organizational structure.

The effort and the time scales required to perform the engineering function will vary widely according to the scale of the enterprise and the novelty and complexity of the concept. In the small firm, there may be only one engineer–designer, who will be surprised to realize that he possesses in some degree all the talents and abilities just described. The intended product may be a slight variant of a well-established and simple design, requiring a minimum of new engineering effort. It may be an entirely new departure from previous practice, requiring research, a number of exploratory design studies (some of which will need to be tested to determine their feasibility), and thorough testing of the final design. This will require the services of a large team of men possessing talent and skill in the specialities referred to. An example of the simple type of product might be an item of domestic equipment. The ultimate example of the second case is the family of Apollo spacecraft.

In both, as in all intermediate cases, there is a common pattern of approach which must always be followed, whether it be by the single design-engineer of the small firm, or by the specialist sections of the larger engineering organization which might be several thousand strong.

The Methodology of Engineering

Man's creative activities, in whatever field, follow a basic pattern which has been recognized for at least 2000 years. Today, it is known as scientific method (cf. page 53, chapter 5). Well-known exponents were Aristotle (384 BC–322 BC), Roger Bacon (1214 AD–1294 AD) and Francis Bacon (1561 AD–1626 AD). It was Francis Bacon who gave greatest prominence to scientific method, and whose name is associated with it. Surprisingly, it is still far too little known and applied by professional managers and engineers, although it provides the guide to all their activities. It will be found to be applied intuitively by most 'little mesthers' —the owners of small firms whose incentive to do the right thing is survival. It is reflected in the Essential Requirements for the Achievement of Reliable Product (page 35) and it forms the basis for the intelligent approach to any problem, whether it be scientific, technological, engineering, or managerial.

Comparing scientific method with the discrete stages in the manu-facturing industrial cycle, we can see the identical logic:

SCIENTIFIC METHOD	MAIN STEPS IN THE ENGINEERING CYCLE
1. Recognition of problem.	1. Determination of needs.
2. Observation of relevant facts.	2. Study of need.
3. Application of knowledge.	3. Design concept.
4. Formulation of hypothesis.	4. Evolved design, i.e. the proposition that the proposed design will satisfy the need.
5. Confirmation of hypothesis by testing.	5. Development testing and finalization of design.
6. Proof of correctness or otherwise of hypothesis, leading to its acceptance or rejection.	6. Manufacture, delivery to user, experience in service.

And the way in which the product behaves in service, the extent to which the hypothesis has been confirmed, completes the cycle and enables the next project to be tackled from a higher plane of knowledge.

It can be accepted that the stages of scientific method are proven and unchangeable. By the same token, it should be accepted that the stages of design and development which have been indicated are essential and vital to the successful outcome of the enterprise.

The main steps in the engineering cycle which have been outlined above can be expanded into the following

Sequence of Engineering Activities
(a) Determination of need.
(b) Study of required performance, conditions of use, and environ-ment.
(c) Project designs.
(d) Study of alternative designs, tests if necessary, adoption of preferred scheme.
(e) Finalization of design scheme.
(f) Calculation of appropriate sizes of components.
(g) Development testing of prototypes.
(h) Value engineering study.
(i) Study of manufacturing feasibility.
(j) Final design.
(k) Preparation of detail drawings, specifications, and standards.

It should go without saying that some of these activities can and will be carried out simultaneously. Some indeed, in the small organization, will be carried out intuitively and instinctively, in the mind of the chief

96

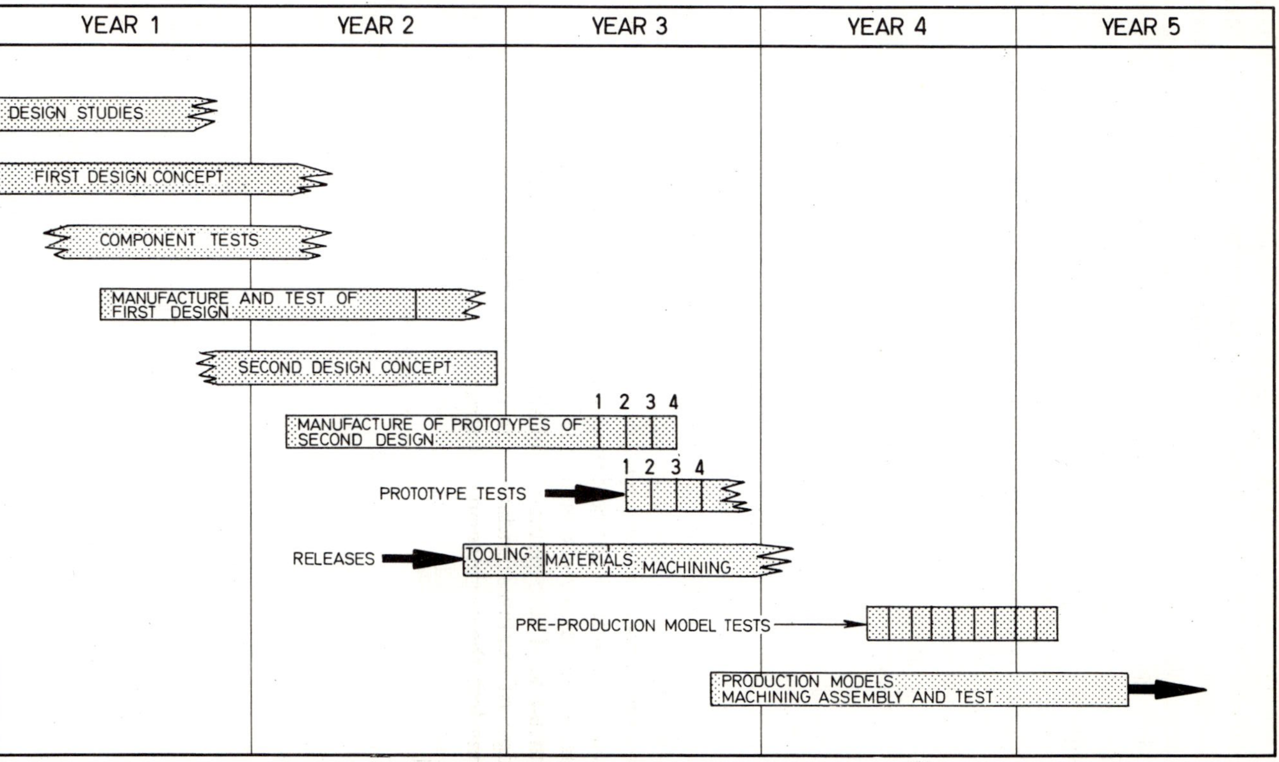

Fig. 7.1. Time Basis of an Advanced Engineering Project.

engineer. Management should recognize its responsibility to understand the importance of these stages, and to ensure that adequate facilities are available. All too often one finds inadequate time and effort allowed for the activities just listed. A common complaint is that there is never enough time or money available for proper design studies, for adequate development testing. The engineer ought not to be compelled to have to request these. It is the responsibility of management to understand the engineer's problems and to make his task as easy as possible. Where the product is complex and the engineering department is a large one, with many problems under review simultaneously, management will be well-advised to enlist the cooperation of the engineers in making a critical path plan. Far from exerting pressures which might be resented, such a plan can help everyone involved in the enterprise by highlighting the problem areas likely to require greatest effort so that the available resources can be employed most effectively.

A management approach is necessary, too, to plan the efficient dovetailing of the design and development processes with those of prototype manufacture, clearance to production, and preparation of production tooling. A typical plan is shown in Fig. 7.1. This is shown for the case of a novel evolutionary product design. The time scale will differ greatly according to the type of project. At the one extreme it will be measured in weeks, at the other, it may be in years.

The activities already mentioned will be familiar to all engineers, in every industry. The pursuit of reliability, however, requires a broader knowledge of certain fundamentals which affect the product after the design has left the engineering department. These are not yet sufficiently well-known, either to engineers or to managers. They are of vital importance to the evolution of satisfactory reliable product, and they will be described before proceeding to a fuller discussion of design and development engineering.

References

1. See:
   (a) Ashby, Eric. *Technology and the Academics.* Macmillan & Co. Ltd., London, 1958.
   (b) Cross, Hardy. *Engineers and Ivory Towers.* McGraw-Hill Book Company Inc., 1952.
   (c) Raistrick, Arthur. *Quakers in Science and Industry.* The Bannisdale Press, London, 1950.
2. Tredgold, Thomas. Institution of Civil Engineers, 1818–1820, London.
3. Eltis, E. M. 'The interaction between the aero-engine industry and growth of air transport.' I.T.A. International Symposium, 25 November 1966, Paris.

8

Engineering for Reliability

It is management's task to organize, guide, and help specialist engineering departments towards the successful achievement of reliable end-product. Specialists are inclined to look upon their own work as of greater importance than that of other specialists, and too often they lose sight of the end-purpose of the enterprise. As Dr Robert Lusser has said:

> Designers (and I am a designer myself) are more interested in achieving high performance, which is fun and glamorous, than in achieving high reliability, which is hard work and entirely unglamorous.[1]

It will help management towards a better and hence more sympathetic understanding of the competing claims of their engineering departments if we consider briefly the fundamentals of reliability. Reliability is a subject which of recent years has tended to become something of a 'mystery' of the exclusive type which was cultivated by alchemists in the Middle Ages. As a matter of fact, difficult as it may be to meet reliability requirements, they can be stated extremely simply.

The Basis of Reliability Achievement
In every problem concerning product reliability, there are only two main factors which must be borne constantly in mind. These are duty and capability.

The simple fact is that if a product is to be completely reliable in use, if it is never to break down, then each and every component of each unit of product must be capable of withstanding the duty which will be imposed upon it.

99

This essential truth has been overlooked by those who have made a narrow speciality of reliability, and who have emphasized that it is a statistical probability, with 100 per cent success only achieved by chance. Yet it is in the aerospace field, where statistics have had their greatest vogue, that we have seen what have probably been the greatest reliability performances to date, the Apollo 11 and 12 moon landings. These took place no more than twelve years after the attempts to prove statistically the near impossibility of achieving 100 per cent reliability. Without question, their success owes much to the scientific and formalized approach to reliability which was pioneered by engineers in the electronics, space, and nuclear industries. But even more vital has been the realization that planned and systematic management of all the multitudinous activities involved was an essential requirement.

It is unfortunate that in the early stages there was little contact between mechanical engineers and the pioneers of reliability engineering in these highly complex fields. The fact that the approach was at first predominantly statistical deterred non-statistical engineers and managers from attacking the subject boldly and resolutely. At the same time, reliability engineers in the electronics and nuclear fields have deprived themselves of a basic approach which would enable them to achieve better reliability, instead of expending considerable effort on the assessment of the degree of unreliability to be expected. The successful doctor diagnoses and cures the pain. He does not waste time measuring its intensity.

In the aircraft and automobile industries in particular, engineers have for a long time had an intuitive understanding of the basic elements of reliability achievement. The vital contribution which has been made by the electronic/space engineers has encouraged mechanical engineers to clarify and to formalize their own thinking. Fortunately, this has led to a simplification of the subject, which should assist engineers and managers in all types of industry to achieve the benefits of increased reliability of their products, at minimum cost.

Duty and Capability It is worth repeating that no product will be 100 per cent reliable unless each and every component in each unit of product is capable of meeting the duty which it will encounter in use.

It will be obvious that no two products will be exactly alike, nor will they have identical strengths and capabilities. Similarly, no two products will receive exactly the same treatment at the hands of the user. One has only to consider one's own experience of motor cars, and of the experiences of one's friends and acquaintances with cars of a given type and make. There is always the likelihood that a component which is slightly below the average of strength or capability will never meet the worst conditions of duty which can be imposed by an individual user. This introduces the element of risk, the *probability* of the statisticians. It makes us realize the fact which must never be forgotten by those concerned with the quality and reliability of products, that there is variability in everything. The capability of components and of the products of which they are constituent parts will vary between a maximum and a minimum value, and the duty will vary similarly.

Moreover, we can never be absolutely certain that the extent of the variations—both of duty and of capability—are known accurately. Unless, therefore, the capability is increased to such an extent that the worst individual specimen will still be adequate for the highest duty which might conceivably be imposed, there will be some risk that the weakest product will be subjected to the most arduous duty and failure of that part will occur. (See Fig. 8.1.)

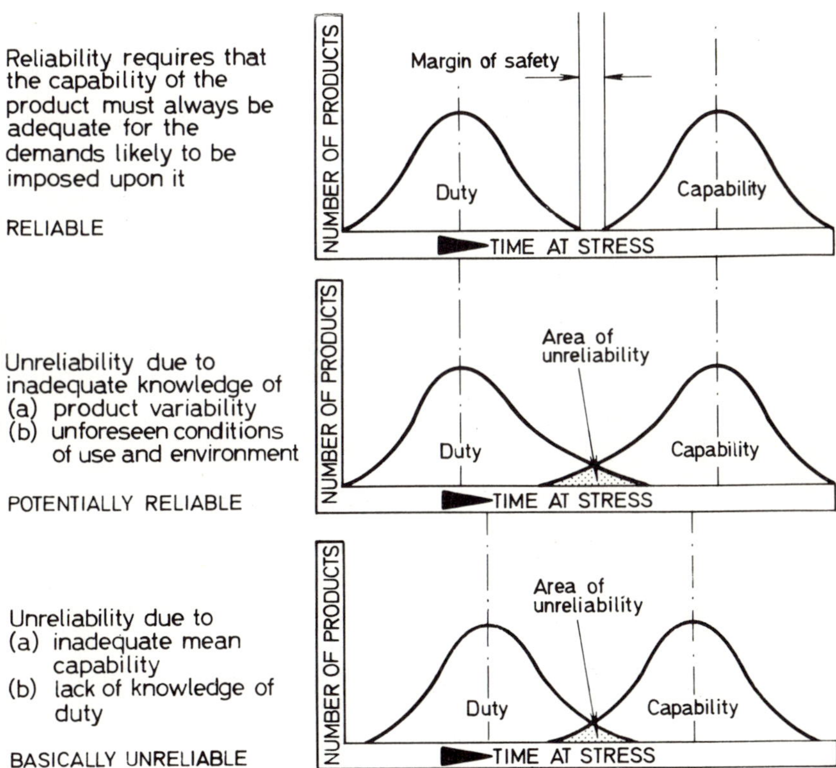

Fig. 8.1. The Basis of Reliability.

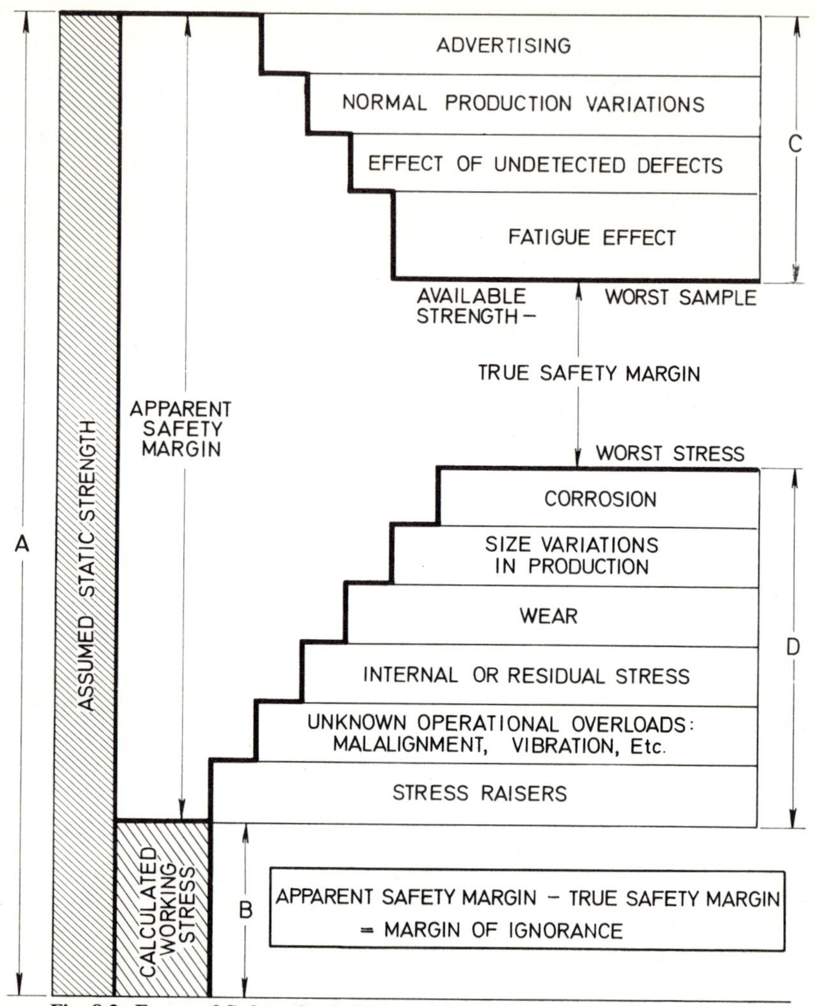

Fig. 8.2. Factor of Safety (So-Called) = A/B. Margin of Ignorance = C + D.

It is important, therefore, that adequate effort should be devoted to the determination and control of the extent of the variation in both duty and capability.

In the majority of design departments even today there persists the concept and practice of designing to a so-called factor of safety. This factor is the ratio of the average strength to the average stress expected to be applied. The acceptable figure for this factor is based upon experience, and the highest factors are to be found in the most traditional and conservative cases. A railway overbridge might have a factor of safety of

8, while a vital component of an aircraft engine might be designed to a factor of safety of 1·2. The term is obviously a misnomer. It disguises the fact that it is indeed a factor of ignorance, while at the same time it is misleading because it engenders a feeling of false security.

Simultaneously and quite independently, two engineers, the author in Britain, and Dr Robert Lusser in the United States, pointed out that the factor of safety was no more than a measure of ignorance. The author introduced the concept of margin of safety, as indicated in the diagram Fig. 8.2.[2]

Lusser, who was working side-by-side with electronic engineers and statisticians, found it necessary to communicate with them in their own language, and he described the same idea in the diagram in Fig. 8.3.[3]

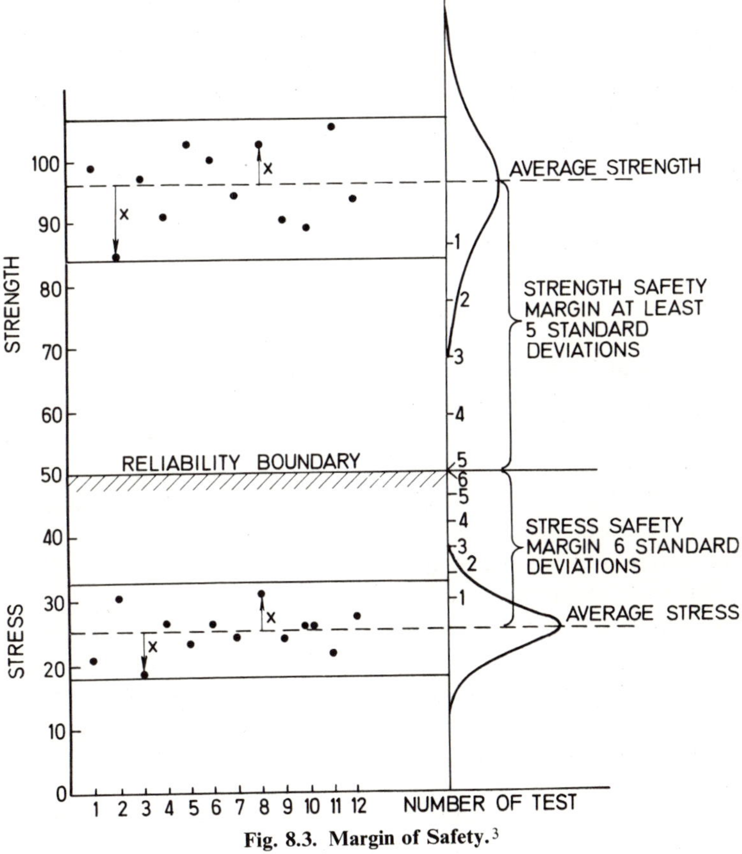

Fig. 8.3. **Margin of Safety.**[3]

Both presentations have their advantages. Figure 8.2 indicates the areas of ignorance which must be studied, by designers, development engineers, and reliability engineers, in order to increase the confidence that there will be a margin of safety, without resorting to too much over-design. Figure 8.3 emphasizes the existence of variability, and the importance of studying and determining, as far as possible, what will be the strength of the weakest product likely to exist in a series production run, and what will be the highest or most arduous duty likely to be imposed.

Testing to Failure A valuable outcome of this approach is the realization that development testing, to determine the capability of a product, is of little value unless it is continued until failure occurs.

All too often development engineers are criticized by their managements for ruining good products by testing them to destruction. Management's criticism is not diminished by the fact that the engineers so evidently enjoy their work. Many designers too, who ought to know better, resent the destruction of their brain children. On consideration, it will be obvious that a successful test of a product, for a given time at a prescribed duty, has done no more than demonstrate that one sample of product has survived that particular test. It cannot establish that the same product would not fail were the test to be continued for a little while longer. It does nothing to reveal, for the benefit of the designer, what is the weakest component—that which would be most likely to fail in service—in the product as a whole.

The Fallacy of the Type Test Quite obviously it will seldom be possible to subject a prototype product to a test of the same duty and duration that it will be expected to withstand in service. A motor car is expected to function reliably and without wearing out for about 100,000 miles. An aircraft engine is expected to run reliably for 10,000 hours or so between overhauls, and to have a total useful life of say 60,000 hours. A steam turbine's useful life might be 30 years, with only brief intervals of shutdown for maintenance and replacement of worn parts.

The art of the engineer is needed to devise tests of much shorter duration, which will reveal to him and to the designer the potential weaknesses in the prototype model. If he is wise, he will continue testing until failure occurs, reducing the time necessary by carefully designed overload tests.

Yet it is quite usual for a customer to accept a new product on the basis of a test of short duration, sometimes of only a few hours, seldom of as much as the 150-hour type test demanded by Air Registration Board and Air Force authorities for aircraft engines and aircraft components.

104

Before subjecting a final product to such an approval test, the engineers concerned with design and development will have sought approval of their management to carry out much more extensive testing. Unfortunately, with the exception once again of aero engines and motor cars, they will usually have been unsuccessful.

Even in these cases, it is customary to select the cheapest of competing products, such as proprietary components or assemblies, on the basis of a type approval test, successfully completed. In the defence field, designers of end-products are often required to make use of Government supplied accessories known as embodiment loan items, in the interests of standardization. Almost invariably these will have been selected on the basis of lowest first cost, subject to the proviso that the product has passed a type test. Unfortunately in many cases the test will have been established for an earlier design of end-product. It will have taken no account of subsequent changes in duty or environment.

Figure 5.14 illustrates the fallacy of the type test as a basis for the demonstration of reliability. Without the background of intelligently planned development testing which, to be of value, must, as already indicated, have led to the failure of one or more components, there is a real risk in allowing a design to proceed to the production stage.

Management's Contribution
The many constituent activities of the design function are dealt with in detail later. At this stage it is hoped that sufficient will have been said to convince management of the need to offer facilities to design and development engineers for obtaining the maximum possible information about the capability of the product and its components, and of the duty likely to be met. There must be sufficient testing to ascertain the variability in capability. This involves:

(a) Work by metallurgists, to determine the scatter or variability in the properties of the materials of construction.
(b) Testing, by development engineers, to determine the scatter in the capability of complete components.
(c) Testing, by quality and reliability engineers, to determine the variability in materials, processes and components produced in large quantities and sometimes in different factories.
(d) The laying down of standards of acceptable quality.
(e) Investigation to determine the extent to which the scatter can be limited by improved processing or by non-destructive examination.

Designers should be given every opportunity to study at first hand the conditions of use and environment. They should be allowed to hold discussions with the user—and by this is meant the engineers who will operate the plant, the soldiers, sailors and airmen who will use military equipment, the housewife who will use domestic aids, or the motorist. Designers should, as far as possible, be allowed to visit the regions in which the equipment will be used, and to see it, or similar equipment, actually in use at the hands of the intended customer.

Examples are far too numerous of costly failures of equipment which have been due to ignorance on the part of the designer of factors with which he could easily have coped, had he known of their existence. The designer himself is seldom in a position to authorize the necessary visits to the eventual customer. It is incumbent upon management to understand the value of these studies, and to see that they are carried out.

References

1. Lusser, Robert. 'The notorious unreliability of complex equipment.' U.S. Army Ordnance Missile Command, September 1956, Redstone Arsenal, Alabama.
2. Nixon, F. *Quality in Engineering Manufacture*. Conference on Technology of Engineering Manufacture, Institution of Mechanical Engineers, London, March, 1958.
3. Lusser, Robert. 'Reliability through safety margins.' U.S. Army Ordnance Missile Command, Redstone Arsenal, Alabama, October, 1958.
4. Nixon, F. 'The joint responsibilities of Government and industrial management.' The Reliability of Service Equipment–Services/Industry Conference. The Institution of Mechanical Engineers, London, February, 1968.

9

The Fundamentals of Fatigue

The vast majority of reliability failures are due to the effects of repeated applications of load. Whether it be a wheelbarrow, or a complicated machine, it is unusual for it to break down the first time it is used, under a single application of load. It is the cumulative effect of stresses repeated many times which causes eventual failure. This is called *fatigue*, which is short for *'metallic fatigue'*, and although it is mentioned more and more frequently in the daily press (vide the Comet aircraft disasters, the Hither Green railway rails, the American F.111 aircraft controls, the *Queen Elizabeth II* turbine blades), it still remains a mystery to many people. This is despite the fact that the phenomenon of metallic fatigue was identified by British engineers as long ago as the eighteen-forties.

In retrospect, it is clear that the fundamental work carried out a few years later by the German, A. Wöhler, took an unfortunate turn. He was a railway engineer, who investigated the causes of failure of railway axles. He designed a machine to test full scale axles in rotating bending, at a speed of 15 rev/min. Becoming interested in the fatigue phenomenon, he developed a machine to use smaller test pieces, which enabled him to speed up the rate of testing and to use less material. Today, the Wöhler rotating bend test machine uses specimens only 0·25 in. diameter or less. Being inexpensive, it has become the standard machine for the great preponderance of research work, and for teaching. Its widespread use has established the study of the stress–life relationship (the *S–N* curve) as the main aim, which was indeed Wöhler's purpose at that time. This has diverted attention from the fact that today we need to be much more concerned with conditions at low lives and at high stress levels.

During the ensuing years, a vast amount of research has been carried out to investigate the causes of fatigue failure, the manner in which a fatigue crack is initiated, how it progresses, and how it responds to applications of load of different magnitudes. While all this is of the greatest potential value in the effort to understand the phenomenon so that fatigue failure can be avoided, it has had the effect of confining most of the work on fatigue to the research laboratory.

A simple pragmatic approach, as was in fact used by Wöhler and his contemporaries in their early experiments, can be of the greatest value. My own experience has confirmed this on very many occasions, and as a result improvements in reliability of great magnitude have been achieved by simple tests based upon approximate assumptions. Because it is felt that these deserve to be much more widely appreciated, they will be described here, so that there can be established a proper basis for the discussion of the design and development activities which will follow.

The Life–Stress Relationship

It is fortunate for us that metals and most other materials of construction behave in a consistent manner. Taking first the most commonly used material, steel, the relationship between the life-to-failure and the magnitude of the stress applied follows the pattern shown in Fig. 9.1.

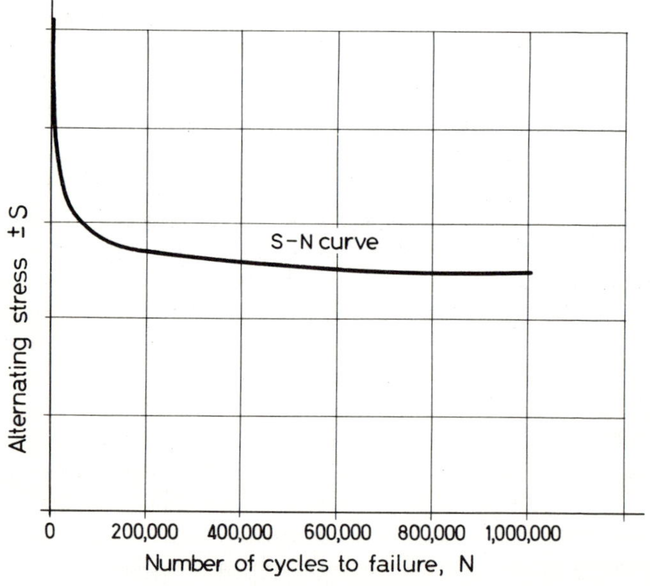

Fig. 9.1. S–N Curve.

108

This is known as the *S–N* curve, where *S* denotes the magnitude of the stress which is applied repeatedly, and *N* the number of times the stress is applied.

A part will fail under one application of load when the stress equals the ultimate which the material is capable of withstanding. If the load is applied oftener, a progressively lower stress will be sufficient to cause failure.

Compared with the work which has been carried out at large numbers of cycles of stress, very little research indeed has been done at low lives, below say 5000 cycles. From about 5000 cycles, however, until about 10 million cycles, there is a continuous fall in the stress which will result in failure. After about 10 million cycles the curve levels off, indicating that stresses below the corresponding figure will not lead to failure, no matter how often they are applied. This stress level is called the fatigue limit, and it holds a peculiar fascination for most workers in this field.

In order to provide a more open scale at higher stresses, it is usual to plot *N* logarithmically. The *S*–log *N* curve, which is the one most commonly seen, is reproduced in Fig. 9.2.

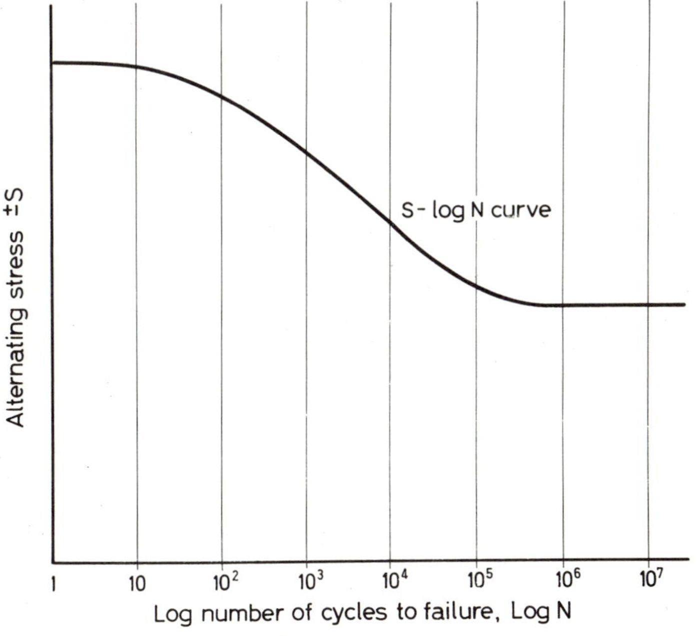

Fig. 9.2. *S–log N* Curve.

For mild steels and steels of medium strength, the fatigue limit is

109

approximately 50 per cent of the ultimate strength. With stronger steels, however, the fatigue limit falls progressively to about 40 per cent of the ultimate strength. Where fatigue conditions prevail, therefore, it is not always advantageous to use a stronger steel. Things might even turn out to be worse than with a weaker steel.

In the case of non-ferrous metals, including aluminium and its alloys, there is no clearly defined fatigue limit, although there is still a distinct bend, or 'knee' in the S–log N curve at the corresponding point. Beyond this, the curve continues to fall slowly, and it can be assumed that if stresses were applied to a test piece for an infinite number of times, the part would fail at near zero stress. This, however, is of no practical significance beyond the fact that it means that at any given stress a component in aluminium has a finite life, whereas in steel, at a sufficiently low stress, i.e. below the fatigue limit, the test piece should, theoretically, last for ever.

Preoccupation with the fatigue limit is due to the fact that Wöhler's original work was in connection with railway wagon axles, which are loaded in rotational bending, one complete stress cycle between positive and negative values occurring at each point of the circumference of the axle for each revolution of the wheel. With a wheel diameter of 3·5 ft, therefore, the axle would be subjected to 10 million cycles of stress in 20,800 miles, or after 77 round trips between London and Derby.

The great impetus to the study of fatigue came from the aero engine designers of the First World War. In their reciprocating engines, where low weight meant high stresses, 10 million cycles of stress would be applied to a connecting rod or a gudgeon pin in about 160 hours of engine running at 2000 rev/min. It was essential, therefore, to know as accurately as possible what was the fatigue limit, and to design the parts so that the applied stress would not reach this value.

These conditions apply to all types of reciprocating engine. Surprisingly, it is not generally appreciated that the majority of today's products are subjected to the full range of fluctuating stress far less frequently.

For example, the steam turbines of a power station suffer one stress cycle from zero when they are at rest to a maximum at their running speed, and back to zero, each time they are started, run, and stopped. Turbines which are shut down once a week will accumulate no more than 1500 cycles in a total life of 30 years. The rotating parts of an aircraft gas turbine engine are subjected, in the same way, to one stress cycle each flight. The total number of cycles will vary according to the purpose and use of the aircraft. A light military trainer, or a short haul airliner might have 6000 cycles of stress annually. A transatlantic airliner, with an annual utilization of 3000 hours, will impose no more than

110

600 cycles. Accordingly it is usual to design the discs of these engines for a maximum life of about 30,000 cycles.

In the case of aircraft, the undercarriage, the flap-operating mechanisms, and many other functioning components suffer no more than two cycles of stress per flight. The disastrous failures of the *Comet* pressurized fuselages in 1954 occurred after a working life of about 9000 hours. With an average flight duration of 3 hours, this meant that the fuselages failed in fatigue after no more than 3000 cycles of applied stress. It was the realization of the fatigue nature of the one stress per flight condition which led to enormous improvements in the reliability of the heat exchanger equipment of the military aircraft of the Royal Air Force in the later stages of the Second World War.

Fatigue has implications of a more personal nature, as biomedical engineers are now realizing. The human hip joint, for example, will be subjected to between 1·5 and 2·0 million applications of load annually. The normal force applied will be about 1·5 times the body weight, but in extreme conditions it may rise to 5 times the body weight. These facts must be taken into account when designing and developing artificial hip joints. In addition, the environmental conditions, especially of the corrosive nature of body fluids, must be allowed for.

In the case of motor cars, whereas the engines will accumulate many millions of stress cycles during their working life, many components—in the transmission and suspension say—will suffer no more than an estimated 12,000 or so cycles of maximum stress range. In these examples it is assumed that no resonant conditions of vibration exist.

With these examples in mind, it is surprising that so little attention has been paid to the low cycle high stress region of the S–$\log N$ curve. Indeed, it will be found that the great majority of the published work records no test results below about 400,000 cycles. When reciprocating aero engines provided the main stimulus for research into the fundamentals of fatigue, it was logical to concentrate upon the fatigue limit and beyond. Unfortunately, a behavioural pattern was established which has persisted, despite the fact that the low cycle region is of far greater importance today. For many years, this has deprived industry of a technique for the establishment of reliable designs which, if more widely known and applied, could probably have prevented some of the catastrophic and expensive failures which have achieved notoriety recently.

Stress-Raisers
Quite early in the history of fatigue it was observed that components tended to fail at oil holes in shafts, at keyways, at shoulders, and corners. Research established that such discontinuities in the otherwise smooth

form of a component can increase the stress, locally, by three times or more in the case of a sharp edged oil hole, for example.[1] It was found that the effect of these stress-raisers could be reduced by providing generous radii in re-entrant corners, and by smoothing off external edges. Surface scratches due to too hasty machining, or to the use of incorrect or insufficiently sharpened tools can be similarly detrimental.

Today, all this is pretty well-known in design and detail drawing offices, and in machine shops. There is still a tendency, however, towards a lack of generosity in providing large fillet radii, especially on castings. On the other hand, although good finish is important, there is evidence that the excessively smooth finishes proposed by research workers are not always strictly necessary on actual working components.

Other stress-raisers, which are especially important when working stresses are high, are defects which may exist in the metal itself. Blow holes or porosity in castings; piping, laps, seams, or roaks in bar or forgings; non-metallic inclusions; fine surface cracks; internal or locked-up stresses in forgings and weldments, all these are capable of raising stress locally to unacceptably high levels. A great deal of progress has been made during recent years in techniques of non-destructive testing to reveal the presence of such undesirable defects. The application of appropriate methods of non-destructive examination is an aid to reliability, because it reduces the areas of ignorance. (See Fig. 8.2.)

Stress-raisers have by far their greatest effect in conditions of fatigue, i.e. where stresses are repeated. Under conditions of static stress, where there is only one single application of load, their effect is very much smaller. This is due partly to the fact that at high static stresses local yielding of the metal will occur, reducing the effective concentration of stress. Although this is well-known, there still persists the widespread practice of carrying out a static proof test in the utterly mistaken notion that it can establish the integrity of a product.

The effect of stress-raisers upon the fatigue strength of test pieces and components is shown in Fig. 9.4. This is presented schematically, because actual examples based on tests extending from one single application of stress to beyond the fatigue limit are rare. This diagram is important as forming the basis for practices of great value to all who wish to achieve reliability of functioning products of any type.

Scatter or Variability

So far, we have presented the S–N curve, and the S–log N curve, as single discrete relationships, indicated by one black line. This is in accordance with the great preponderance of published data on fatigue, the results of countless research investigations.

112

These suffer from the serious defect that they assume that all samples of material behave in the same way. They ignore completely the variability of materials, of test pieces, of components, which must always be borne in mind by the engineer who is concerned with the achievement of reliability. Unfortunately, it is a matter of sober and regrettable fact that variations in the quality of materials, which still elude our ability to measure and control them, cause variations in fatigue strength and fatigue life which are quite considerable.

Perhaps because research workers have been more concerned with the establishment of relationships between stress and life, S and N, or possibly because the truth would require them to investigate the arduous

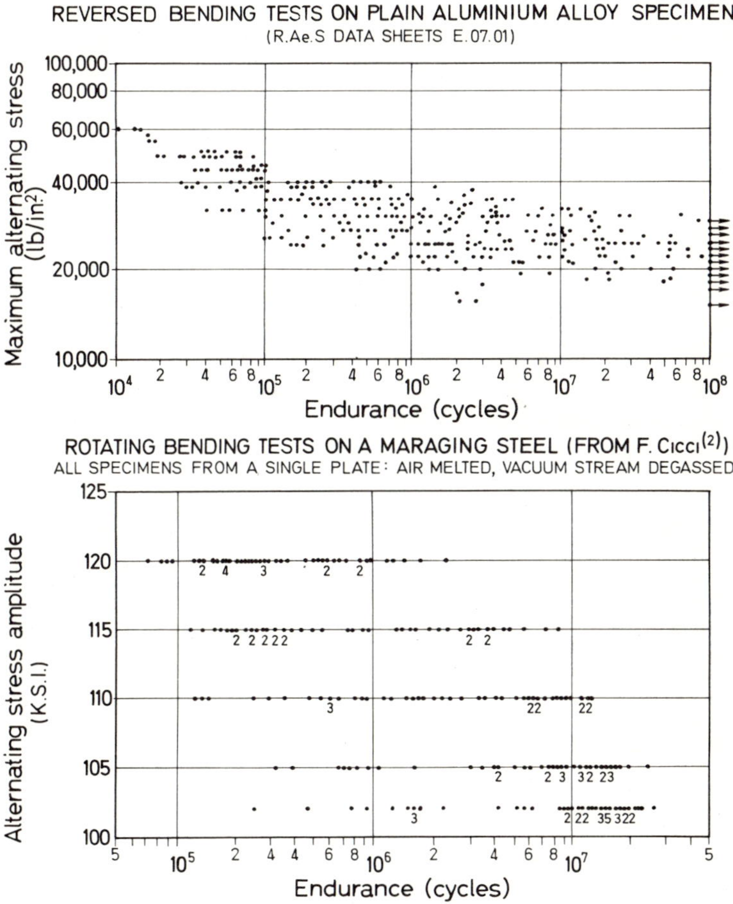

Fig. 9.3. **Examples of Scatter in Fatigue Life and Fatigue Strength.**

and unglamorous aspects of the pursuit of reliability, they have, with a few notable exceptions, closed their eyes to the existence of scatter in the results of tests of fatigue strength and life. Indeed, more than one professor of mechanical engineering has confessed, when challenged, that he has disregarded test results which did not fall on the anticipated and tidy 'normal' line. The information which was not recorded could have been of far greater value to the designer than the precise form of the nominal curve which the researchers were at pains to establish.

Figure 9.3 shows examples from the relatively sparse published data indicating the scatter in fatigue characteristics. This scatter is of such magnitude as to occasion more than a little surprise that it has been ignored for so long. The variation in fatigue life which can be expected, in test pieces from laboratory specimens and from actual components, may be as much as 100 to 1 between the maximum and the minimum life. This is on the basis of a scatter of $\pm\, 3\,\sigma$, that is, assuming that 99·7 per cent of the total population will come within these values.

In the case of fatigue strength, as distinct from fatigue life, the variation between maximum and minimum will be seen to be of the order of 1·5 : 1.

It is important to realize how misleading is the data which is provided to designers. It is quite astonishing that standard specifications for materials ignore the fundamental fact of variability. The extent of the variability could easily be defined by quoting the *mean* and the *range* of figures to be expected in a batch of material. This was pointed out more than 30 years ago by Professor Egon Pearson,[3] and later by W. T. O'Dea[4] and by Kinzel,[5] but action by national standards institutions and by material suppliers is still awaited.

Non-Destructive Testing
Granted that information has been obtained on the extent of the scatter in fatigue properties, the decision must be made as to whether the lowest figures must be accepted as inevitable, and the design based accordingly, or whether a minimum standard at a higher life can be laid down which can be maintained by eliminating the lower end of the scatter by non-destructive testing to identify the worst components.

This is an area of activity which has so far received little attention outside the aero engine and nuclear fields. Fortunately, it is far less difficult to establish acceptable standards than might be thought. This can be done by taking advantage of the characteristics of the *S*–log *N* curve at its higher (low life) end.

An Engineering Approach
Determining the Extent of Scatter Information of inestimable value has
114

been obtained, quickly and economically, by making use of the standard form of the *S*–log *N* relationship.

Figure 9.4 indicates, schematically, the fact that there will be scatter in the life and strength of individual samples. We need to determine, as quickly as possible, the extent of this scatter. To follow the usual practice of determining the scatter in fatigue strength necessitates carrying out tests totalling hundreds of millions of cycles, because at least 7 and preferably as many as 30 test results are necessary to enable a statistically viable estimate to be made of the total scatter. Moreover, since the fatigue curve in this region beyond the fatigue limit is open-ended, it is difficult to control the test programme. The amount of testing might total at least 2000 hours and this fact has already been a sufficient deterrent to the determination of such data.

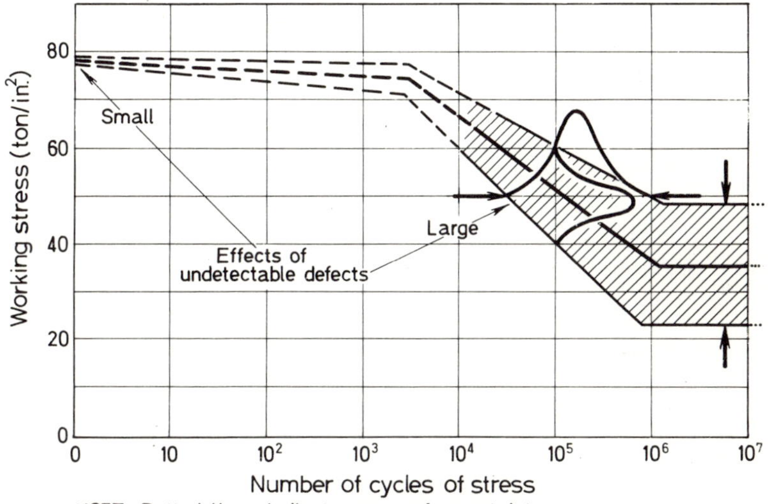

Fig. 9.4. Ineffectiveness of Static Test, Effectiveness of Cyclic Test, to Indicate Soundness of Product.

If, however, we determine fatigue *life* rather than fatigue *strength*, by working in the high stress low cycle area of the curve, things become much more feasible. By carrying out tests on specimens at a high level of stress, approaching the 0·1 per cent proof stress, conditions will not be significantly different from those existing at lower stresses. At this stress level, the test pieces will have a finite life and it will be possible to test even the desirable 30 specimens for a total test duration of about 1 million cycles, or from 8 to 10 hours' testing at 2000 cycles per minute.

As we have seen already, the great majority of today's products do

115

have finite lives. If, however, a particular component, e.g. a crankshaft, connecting rod, gudgeon pin of a reciprocating engine is under review, the scatter in fatigue strength beyond the fatigue limit can be estimated from the scatter in fatigue life. The slope of the S–log N curve does not vary greatly for a given metal, and a sufficiently accurate estimate for practical purposes can be obtained by sketching in a curve based upon the predetermined life-scatter at a given stress.

Standards of Acceptable Quality Short life tests of this kind have another great advantage. They sort out for us the specimens which are better than average, from those which are worse than average. It is then possible for the metallurgist to compare these to try to detect the significant factors influencing goodness and badness. This provides, too, material for the development of methods of non-destructive examination. If a method of non-destructive testing can be evolved which can with confidence predetermine the quality of a specimen, then it becomes possible to specify a minimum standard of acceptable quality and to reduce the scatter in the fatigue life which must be catered for in the design.

This approach has been of great value not only in improving reliability, but in providing a challenge to the development of improved methods of non-destructive testing. Indeed, the engineers carrying out the fatigue tests, and those working on non-destructive testing, become mutually self-fertilizing, each group challenging the other to greater heights of achievement.

Simple Equipment The method described was developed by quality engineers, using test pieces taken from actual components. Because it is based upon low-life testing, it has been found possible to use simple home-made test rigs, operating at moderate speeds, which can be constructed quite cheaply. Yet these have given results comparable with those obtained from the most advanced and expensive fatigue machines. Figure 9.5 compares test results obtained on a simple constant deflection test machine operated by a $\frac{1}{4}$ hp electric motor, used in a workshop, with those from a Vibrophore machine, used in a research laboratory on similar specimens.

It is to be expected that research workers who have investigated the effects of rate of cycling and of overstressing by prior damage will object that this pragmatic approach will produce misleading results. More than 25 years' experience, with a wide variety of components and many different types of construction including castings, forgings, machined forgings, sheet metal fabrications soft soldered, brazed, and welded, in brass, copper, mild steel, stainless steel, austenitic steel, nimonic alloys, light alloys, and titanium alloys, has however not yet

116

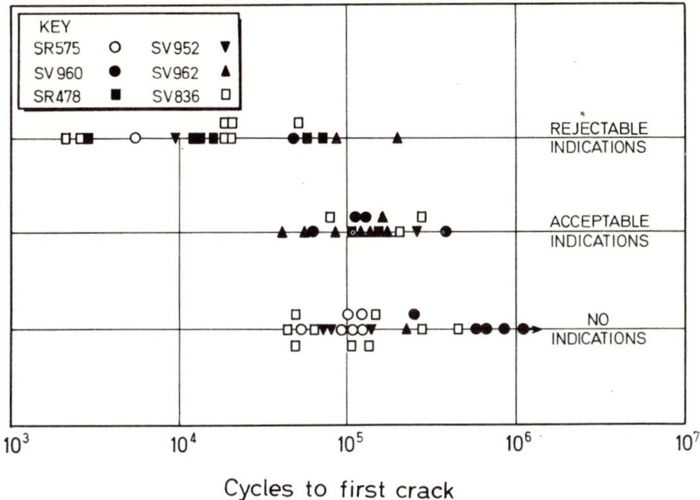

RESULTS OF TENSION FATIGUE TESTS ON 0.45in. WIDE STRIPS
CUT FROM DIAPHRAGMS AND TESTED ON AMSLER
VIBROPHORE MACHINE

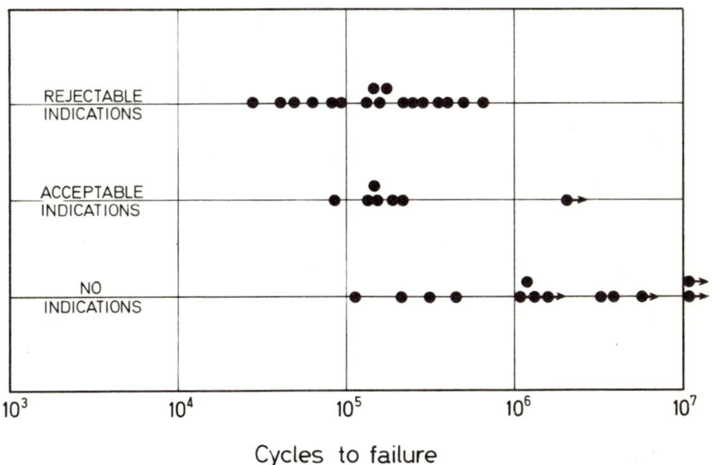

Fig. 9.5. Comparison of Results obtained from a Simple Rig and a Laboratory Machine.

117

produced spurious or doubtful results. On the contrary, it has been possible on many occasions to reproduce identically the type of failure occurring in the full scale component, even though the actual stress conditions have not closely simulated those of the engine proper.[6] As a result great improvements in reliability have been achieved.

It is important to distinguish between this work, which falls within the sphere of activities of quality engineering and reliability engineering, and that of the development engineer. It is the task of the development engineer to establish the viability of a design. The quality/reliability engineer is concerned with detecting and eliminating the causes of trouble in production units, which are due to variations in components produced in large numbers.

Proof Testing Another valuable aspect of this approach is the facility which it offers of proving components the integrity of which cannot be established by any known method of non-destructive testing.

Such components include heat exchanger equipment, such as coolant radiators and oil coolers and heat exchangers in air-conditioning plant. These are usually constructed of thin foil in copper, brass, steel or light alloy, joined by soft soldering, brazing, or welding. They include myriads of joints, which cannot be examined.

Another family of components is that having welded or brazed joints. It is not generally realized that in the state of the present art of inspection, while it may be possible to detect a defective weld or brazed joint, it is not possible to be assured that the joint is a good one. X-ray examination can reveal defects. It cannot confirm the existence of good fusion.

A great deal of what has been written in this chapter applies as much to complete functioning components as it does to test pieces or single discrete details. A defective valve in a pneumatic or hydraulic jack, a defective limit switch in an electric motor, can act as a life reducer (i.e. the equivalent of a strength reducer) just as can a hairline crack in a metallic component.

In the examples mentioned, it has been found that the stress-raising or life shortening effect of the typical defect is considerable relative to the normal scatter to be expected in the population of sound products. The situation can be illustrated by Fig. 9.6.[6] The effect of the high stress-raiser is to produce a virtual separation between 'good' and 'bad' families of components. This separation has been found to exist in a wide variety of products. There will not, however, be this separation in the case of non-metallic inclusions in forgings. Here, the effect is continuously deleterious as between the less and the more harmful inclusions so that there is simply one wide scatter band.

118

It must be emphasized, too, that the author's experience has not included heavy weldments. In no case should the method be applied until
it has been established that the situation of Fig. 9.6 does in fact exist.
Then, by imposing a test of duration, typically of 2000 cycles on every
component produced, there is a high probability that all defective ones
will fail while the life of good components will be reduced by no more
than about 2 per cent. This is a small price to pay for the assured integrity
of the products which pass the test.

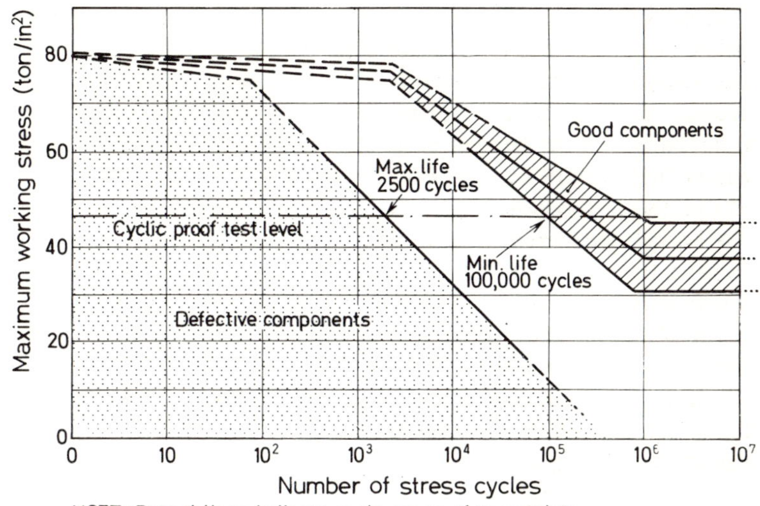

Fig. 9.6. The Basis of Cyclic Proof Testing.[6]

The approach which has been described is far removed from that of
the research worker. Hence, no manager should accept the excuse that
such testing cannot be carried out because research facilities do not
exist, or because the laboratory is too busy. The tests are basically for
engineers, using the minimum of easily built-up equipment. Their value
has been established so convincingly that it is a matter of surprise that
they have not been more widely adopted.

Vibration Troubles

No mention has been made of the effects of high speed vibration,
because resonant vibration is an unwanted condition which must be
detected and eliminated from the design if it is to function satisfactorily.

We all know that soldiers break step when crossing a bridge. The
rhythmic marching tread of a number of men provides a periodic

119

exciting force. If the frequency of the step should coincide with the natural frequency of the bridge, or a multiple or sub-multiple of it, then vibration would be set up which would increase in amplitude with each successive tread. Even though the force applied might be small compared with the strength of the bridge, a sufficient number of applications of the force could cause the bridge to fail.

One person crossing a plank bridge over a stream can easily set up resonance. Usually he has hurried off the bridge before he has been thrown off by losing his balance.

In the same way quite small periodic forces in machines can cause failure of components which have an appropriate natural frequency so that resonance occurs. The tooth contacts between mating gears may produce no more than a buzzing, which may be detected as a faint vibration felt by the finger tips. Should its frequency coincide with the natural frequency of a component, then serious trouble may be anticipated with that component.

Valve springs surge in tune with the harmonics of the cam, crankshafts oscillate torsionally under the influence of periodic explosion forces. These conditions have long been known, and today they are coped with.

The blades of a steam turbine may be subject to excitation due to their frequent passage past the steam nozzles; those of a gas turbine due to variations in gas flow intensity. The natural frequency of a steam turbine blade may be of the order of 1000 cycles per second. It will vibrate at this frequency when the exciting force is in resonance. Only about 3 hours of accumulated running time at speeds producing resonance will be necessary to impose 10 million cycles of stress. These vibration stresses, which will alternate between positive and negative values, will be superimposed upon the centrifugal stresses due to rotation of the turbine and the bending stresses due to the steam. These stresses will be unvarying at a given running condition, but it is known that the higher this steady stress, the lower the resonant stresses that can be withstood without failure. There is a relationship between mean stress and the range of fluctuating stress which is approximately as shown in the Goodman diagram. (See Fig. 9.7.)

The trouble experienced with the turbines of *Queen Elizabeth II* was, in fact, due to resonance between the blades of certain stages of the turbine and the nozzle frequency. The trouble was overcome by modifying the blades to lower the running stresses, and by lacing them together in batches to change their frequency.[7]

Often, the critical period extends over only a narrow range of operating speed. A detailed and painstaking search may, therefore, be necessary to ensure that resonance does not in fact exist. Quite small dimen-

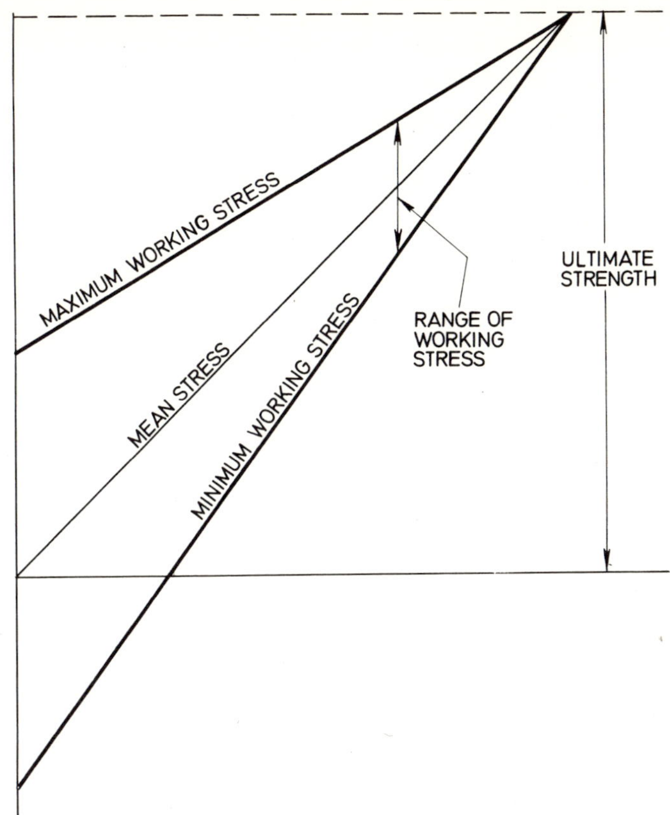

Fig. 9.7. Goodman Diagram, Showing Relation between Range of Stress and Mean Stress.

sional changes can alter the natural frequency of a part and put it at hazard.

The possibility of resonance existing in a design should never be overlooked. The consequences can be dire, and failure to carry out thorough tests, especially on expensive or critical products, has resulted in extremely costly breakdowns.

References

1. Peterson, R. E. *Stress Concentration Design Factors*. Chapman & Hall, London, 1953.
2. Cicci, F. 'An Investigation of the Statistical Distribution of Constant Amplitude Endurances for a Maraging Steel.' UTIUS Tech. Note No. 73, July, 1964.
3. Pearson, E. S. 'The Application of Statistical Methods to Industrial Standardization and Quality Control.' BS 600, 1935. British Standards Institution, London.

4. O'Dea, W. T. 'A statistical examination of specifications for the mechanical testing of line insulators.' *J. Inst. Electrical Engineers,* Vol. 83, No. 501, London, September, 1938.
5. Kinzel, Augustus B. 'Specifications?!' H. W. Gillett Memorial Lecture, 1961. American Society for Testing and Materials, Philadelphia, U.S.A.
6. For a fuller discussion see Nixon, F. 'Testing for satisfactory life.' Symposium on Relation of Testing and Service Performance, American Society for Testing and Materials, Atlantic City, N.J., U.S.A., June, 1966.
7. Fleeting, R. and Coats, R. 'Blade failures in the H.P. turbines of RMS *Queen Elizabeth 2* and their rectification.' The Institute of Marine Engineers, 28 October 1969, London.

10

From Concept to Prototype

Management may perhaps be forgiven for failing to appreciate the importance of the design phase in the evolution of a satisfactory product. The literature shows that technical and professional individuals and organizations find difficulty in defining the function of design, indeed in defining design itself. In different organizations, a man carrying out the same duties might be described variously as an engineering designer, as a designer, or as a draughtsman. In other firms, the designer may in fact be a stylist, or an 'industrial designer'. Added to this is the fact that good engineering design calls for a great deal of judgement and intuition, and much of it is an art which cannot easily be described to non-engineers. Finally, the design department is often regarded as non-productive, as an 'indirect' drain on the company finances. Managers who are accustomed to assessing performance in terms of output in a given time, for a given cost, find difficulty in evaluating design effort which at best may result in a collection of drawings, at worst in the abandoning of what might have appeared, to the sales director say, to be a promising project.

So much for design in the mechanical, structural, civil, chemical, and electrical fields, where the product is a complex of many components, each one of which is called upon to play a part, dynamic or static, which is contributory to the functioning of the whole device. The product, whether it be a prime mover, a machine, or a processing plant, can be compared to the human body, with its skeletal framework, its muscles, heart pump, lubrication system, heat exchanger, and means of converting fuel into energy. In electronic engineering, the components are more nearly analogous to the cells of the brain. They receive and transmit signals in response to stimuli, and form a complex system of actions and interactions. In this case, design is the word used to describe the arrangement of the system, its logic. This kind of design, however, is capable of being treated much more rigorously as a rigid scientific discipline than is, up to the present time at any rate, mechanical design. At the same time electronic devices still require to operate through the medium of mechanical, structural, and electrical components. To this extent, they are subject to the art rather than to the science of design.

Design in the Doldrums
With such a diversity of tasks, with art and intuition on the one hand, and the mysteries of advanced science on the other, with a confusion of semantics, it is small wonder that designers often fail to justify their existence in the eyes of their management. They are unable to win the status and the awards which they feel to be their due. They are not accorded the time nor the facilities necessary to enable them to do their job to their own satisfaction. The unfortunate consequence of this lack of appreciation of the important role of the designer is that it becomes the ambition of many able men to 'get off the drawing board'. Those who do not succeed in doing so are driven to adopt a defensive attitude which isolates them from the rest of the organization in which they work, and an already unsatisfactory situation is made worse.

The too frequent occurrence of catastrophically expensive failures of large one-off equipments is ample support for the general impression that too little heed has been paid to the Feilden Report. It is hoped that the following account of the design function and the basic pattern of design activities will help management towards a better understanding of the way in which the designer works, why certain expenditure is justified, and perhaps most important of all, how to integrate the design departments more effectively into the organization as a whole.

The designer and the manager both work in essentially the same way, applying, by their art, the totality of existing relevant knowledge—the science of their subject. It might have been thought, therefore, that there would have been a better understanding of each other's problems than does in fact exist.

The Responsibilities of the Designer
Despite the low status which is so often accorded to the design activity, the fact remains that the designer bears a much greater responsibility for the success of the whole enterprise than does anyone else in the organization.

We have seen already that reliable performance depends predominantly upon design. The volume of sales, the competitiveness of the product,

124

its profitability, are all determined by the designer above all others. Should the quality of the design be low, or should the product not meet the customer's requirements adequately, no sales effort, on whatever scale it might be mounted, will have a lasting effect. Indeed, to 'plug' a product which is less than adequate for the customer's needs approaches confidence trickery, and a disappointed market will remember this. Competitiveness requires that the product shall be of equal or better value to the customer than he can obtain elsewhere. Value, as we have seen, depends upon quality and reliability and price. Profitability depends upon cost related to the price which can be demanded. Here, despite the efforts which should be made by production engineers to reduce manufacturing costs, their work is circumscribed by the basic pattern of the design, which determines the cost bracket within which the product lies. The price which can be obtained depends to a great extent upon the customer satisfaction which is offered.

A product design represents the designer's opinion of what he considers will be successful in satisfying the needs of a particular market. Although there will have been a corporate decision on the basic type of product and on the sector of the market to be attacked, the interpretation of that decision, the translation of intention into hardware, is the responsibility of the designer. It should be obvious that this activity is of fundamental importance to the success, indeed the survival of the company. Yet how often is the designer given adequate time and facilities—above all adequate time—to be able to devote the careful thought necessary to arrive at the best decision!

The weight of responsibility which falls upon the designer will vary widely according to the type of product, and upon its state of maturity. This variation is possibly an additional reason why it is difficult to arouse the interest of management by generalizations regarding the importance of design.

The Life-Cycle of a Design Concept R. J. Dain has identified seven ages in the total life-cycle of a product design.[1] Illustrated by topical examples, these are:

1. INVENTION A new concept is conceived, e.g. the Wankel engine, through-the-lens exposure metering for cameras.
2. DEVELOPMENT OF THE IDEA Conversion into practical reality, e.g. Wankel engines for motor cars.
3. INITIAL COMMERCIAL INTRODUCTION The launching of small quantities, in the beginning at high prices, e.g. custom-built sports cars of high performance.

4. TECHNICAL IMPROVEMENTS Development to meet customers' demands, e.g. the current move to design safer motor cars.

5. INCREASED COMPETITIVENESS Refinement of design to improve value, e.g. by enhanced quality of design, better reliability, lower costs; e.g. the mini motor car.

6. INCREASED PROFITABILITY Cultivation of mass market, improved detail design for lower production costs, application of value engineering, more economical methods of manufacture by new plant and equipment, better planning.

7. REPLACEMENT BY A NEW DESIGN For example, the highly developed car ferry replaced by the hovercraft, the petrol-engined motor car replaced by the small electric town vehicle for commuters.

Only the largest organizations will find themselves engaged in the whole range of activities covering this total life-cycle. Even then different phases will usually be confined to different design departments. In general, most companies will have concurrent activities in no more than three or four of the phases just described.

At any given time an established organization will find itself with its main product or products in one of the later stages described by Dain. It will be well advised to have under active consideration new products in the earlier stages, and to ensure that there is a continued progression from one stage to the next. The type of design activity required at each stage will vary in degree, but in every case the designer must be responsible for keeping abreast of technical developments in fields ancillary to his own.

It should go without saying that the designer should be aware of changing needs in the market, of new products being evolved by competing firms. He should know of new materials, new manufacturing processes, new methods of inspection. One reason for the growing consciousness of the importance of the design function is the fact that demands are increasing concurrently in so many areas. The greatly expanded consumer market has led to intense competition in the field of consumer durables, and the customer is becoming increasingly educated and critical. The restraints on Government spending on defence which have had to be made to balance the increased spending by and for the domestic consumer have led to demands for enhanced value, increased reliability and life of military equipment. The greatly increased performance of land-vehicles, of aircraft, of warships means that the designer must strive continuously to keep up-to-date his knowledge of new technical developments.

All this is necessary if a company is to remain in business. We can all

126

call to mind instances of firms which, after enjoying a long period of prosperity due to a high reputation for a good design, have declined and failed because they have not kept abreast of their competitors.

The vital role of the designer should now be manifest. Yet woe betide the designer whose product suffers a breakdown. The adverse publicity might well be worldwide. Surely, to counterbalance this there should be a better realization of the heavy load which falls upon the designer, and he should be given credit and praise for work well done. Management should bear this in mind.

Prince Philip, in his speech launching National Quality and Reliability Year, said:

> A good reputation for well-designed goods or components, fit for the purpose, which don't fail or break down, is the criterion for certain success. A bad reputation is a very costly luxury which this Nation cannot afford.

The Function of Design

The designer's task is to create a specification which will make possible the manufacture of a product meeting the specified requirements as economically as possible.

In order to discharge his task satisfactorily, the designer must first know what are the needs, the requirements, or the expectations of the intended customer. Whether they are needs or expectations will depend upon the type of customer. The large organization which by the nature of its purpose must determine its needs for itself—a defence department, a large chemical manufacturer, a large manufacturing organization requiring new plant or equipment, an airline—will itself define the performance, the operating cost, the cost of ownership, the delivery dates, which must be met. The demand of the individual customer, on the other hand, will be much more in the nature of a reaction to the supply of what is available. In this case, the needs of the market will be defined in the first instance by the market research department of the organization which achieves a breakthrough and hence becomes a trend-setter. Competing organizations must make haste to diminish the lead thus established. In either event, the designer finds himself presented with a broad expression of need. It is his function to visualize the need, to see in his mind's eye the conditions in which the product will be used— and abused—and to synthesize a form of product which will satisfy the requirements as he sees them.

The more successful designer is he who possesses creative ability. This means that he is able to evolve, at first perhaps intuitively, a form

127

of product likely to be able to meet the specified performance. It is then necessary to confirm that the concept will be able to meet the requirements of performance, strength, safety, reliability and so on. Since these characteristics will probably be in conflict with each other, it is not unlikely that the designer will have to decide upon a compromise—to 'trade off' some part of the desired minimum weight, for instance, in exchange for adequate structural safety. This second phase of design, which in effect adapts the initial concept to practical reality, can be dealt with by engineers possessing the necessary technical knowledge of stressing, of dynamics and strength of materials, but not necessarily having creative ability. Depending upon the novelty of the design, more or less testing will be necessary to confirm the assumption made by the designer.

When the design has been finalized, it must be defined and specified, usually by means of drawings, specifications, and instructions.

During the stages of creation of specification, finalization of specification, definition and communication, it will help the designer to bear in mind the main areas of requirement, which will vary according to the type of product and its end-purpose. These provide the designer with a broad guide which can be broken down into greater detail.

The following list of the more important characteristics required by the Navy, in its ships, machinery, and equipment is due to Rear-Admiral W. A. Haynes.[2] It would require only little change of emphasis on individual items in this list for it to be applicable to any kind of product.

> Performance.
> Reliability.
> Robustness.
> Maintainability and repairability.
> Ease of operation.
> Silence.
> Safety.
> Compactness and lightness.
> Habitability and appearance.

The activities involved in arriving at a specification of product to meet requirements as indicated above are primarily the concern of the engineering department—the designers and the development engineers.

Optimum value, which involves manufacturing cost, requires that the designer should consult engineers concerned with the assessment of value, with the planning of manufacturing methods, with manufacture itself, with quality assurance, and quality control. Since there is often found to exist the tendency towards departmental isolationism which

128

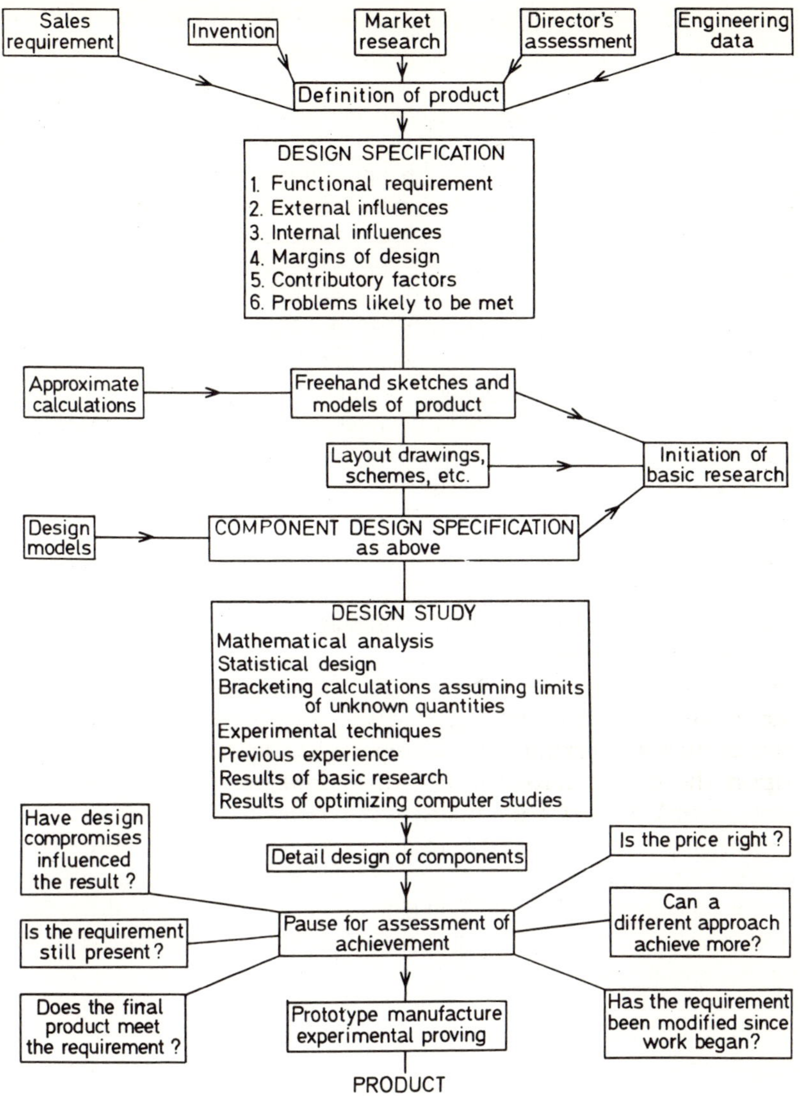

Fig. 10.1. Flow Chart for Design Process.[1]

has already been mentioned, it is incumbent upon management to ensure that means exist for bringing about the necessary cooperation, and for seeing that it is effective.

The essential interdependence of design with the other activities of the organization is shown in Fig. 10.1. This will provide a key to the ensuing

129

account of the important specialist activities which comprise the whole of effective design.

The Design Process

When management has reached a decision on the type of product needed, the designer has in effect received his instructions to proceed with the evolution of a design of product which will fulfil the hopes and aspirations of the enterprise by satisfying the intended market.

Whether they take place within the brain of the chief executive of the small business who doubles the role of chief designer, or whether they require the services of many engineers and designers staffing a complex of specialist departments, there is a logical sequence of stages of thinking, decision, and action. Their importance and their function must be recognized and understood if they are to be managed efficiently and implemented in appropriate measure.

Determination of Need The need may arise from the simple personal desire on the part of the proprietor of a company to make something which he wants to make, to satisfy his creative instinct. Many important new concepts have arisen in this way.

It may arise from a desperate need to meet the competition from another more forward-looking enterprise which is winning the market. This is a common determinant of company policy. Many managers will look upon the determination of need as market research. They should, however, guard against the tendency to regard this as a separate independent activity, concerned mainly with the study of the size of market for a particular product and the price which the market will stand. Of at least equal importance should be the consideration of the product itself, of the feasibility of evolving a product for the appropriate market at a price which will allow it to be satisfactory, competitive, and profitable. This means that the designer should be involved from the outset. Merely to instruct the designer of a product to meet broadly-stated requirements and firmly-stated targets of delivery date and cost may lead to satisfactory sales in the short-term. Such a plan is not likely to lead to long-term success, mainly because insufficient time will be allowed for proper design and development.

In other cases, a demand may arise for plant or equipment to help a large organization to improve its situation. As an example, a chemical firm may decide that it will be more economical to produce basic materials in much larger quantities by large-scale plant continuously operating. A shipping company may need a new passenger liner to attract more custom, or a much larger cargo carrier to improve the economy of operation. An airline may have to keep pace with competition, or ahead of it, by procuring faster and more economical and quieter aircraft.

Perhaps an even greater incentive comes from the needs of defence. *Pace* those who advocate universal disarmament, man's history is one of continuous evolution of means of offence and defence. In every country, there are defence departments conducting research into the strategic and tactical problems arising out of their particular situation. In most countries, the expenditure on defence is a large proportion of the Gross National Product. Hence the demands imposed upon industry, for defence hardware of all kinds, form an important part of a country's manufacturing economy. Also of great importance are the financial assistance given to industry for study and research into specific problems, and the 'spin-offs', the new techniques and technologies which derive from this work and which have on occasion founded whole new industries.

Study of Need The expression of a need, whether it comes from within the organization or from a potential customer, must be clearly defined if the designer is to have a real opportunity of satisfying that need. This phase is probably the most important of all. More trouble arises through failure to communicate to the designer precisely what is required, what performance is expected of a product, how it is likely to be used and abused than from any other cause. We have already seen that inadequacy of design accounts for the greater part of unreliability. It is not the ability of the designer which is in question, but rather the adequacy of the specification of requirements with which he is provided, and the facilities which are afforded to him to find things out for himself.

He needs to know what are the conditions of use and environment to which the product will be subjected. Is low weight important, implying high stress levels? Is rigidity of greater importance? In what climatic conditions will the product be used—hot and dry, hot and humid, cold and dry, cold and wet, dusty, muddy, salty, or any combination of these? Are corrosive atmospheres likely to be encountered? What is the degree of skill, or potential for being trained, of those who will operate the equipment?

Large organizations, nationalized industries, industrial firms, defence forces, large retailers, all these have technical departments capable of specifying in some detail their requirements, and the anticipated conditions of use. In some cases (some armed services are among the worst offenders), however, so many different departments may be involved that time which would have been valuable to the designer is denied to him.

Conflicting demands will come from those departments concerned with offensive weapons, defensive armour, size, speed, range, delivery date, initial cost, reliability, and cost of ownership (i.e. life-cycle cost). Sometimes several years may elapse before a mutually satisfactory compromise has been reached.

Although the reasons are understandable, especially in times of financial stringency, here is another indication of failure to appreciate the importance of the design function, which as usual is left until last, and is then expected to achieve good results in an impossibly short time.

Management should realize that this phase of the design cycle is crucial. The designer should be given the opportunity to take part in the discussion of need with the potential customer. Whenever possible, preliminary design studies should be authorized while these talks are proceeding. These would help to determine the feasibility of the project, and in appropriate cases this work should be financed by the customer.

When he does not already have direct experience, the designer should be sent into the field, to study the conditions for himself. After all, it is he who will be committing his firm, and the customer. The cost of a short trip overseas to see at first hand the intended conditions of use is a small price to pay for the better quality of design which should result.

A Planned Approach The designer should make it his first task to write down his version of what he thinks is required. He will do this under headings such as the following:

1. FUNCTIONAL REQUIREMENTS Performance, environment, conditions of use, expected reliability, serviceability, maintainability, repairability.
2. SPECIAL CONDITIONS OF DUTY Operating temperatures, vibration, shock, radiation (as in space), corrosion, dust, humidity.
3. SPECIAL DESIGN PROBLEMS High stress levels, materials stipulated by the customer, lubrication.
4. MARGIN OF SAFETY Risk of overloads, capacity for uprating in the future.
5. COST AND VALUE FACTORS Acceptable price, rate of production, total life cost, extension to other fields of application.
6. POSSIBLE UNKNOWNS Anticipated hazards and their probable effects, causes, and avoidance.

Such a list can be used as the basis for questions to be posed to the customer, in order to improve the specification of requirements. It can be used by the customer to improve the initial specification, and it can form a useful guide for the next phase of design.

Preparing the Design No matter what the level of sophistication of the product, the designer will approach his task in the same way. He will see in his mind's eye possible ways of meeting the need, and he will usually start by setting down his ideas on paper, in the form of preliminary sketches.

In many cases, the designer's experience will enable him to decide appropriate proportions and dimensions. Provided that established practice is followed reasonably closely, it is highly probable that the product will be satisfactory. Because the designer is so often under pressure to complete his work too quickly, however, even with relatively simple products there is the risk that he will have been compelled to concentrate upon the more critical components. In consequence, it is often found that designs fail, not because a part has been stressed too highly, but because some relatively minor point has been overlooked.

It has been the author's experience that far more often than not trouble with a product in service has been found to be due, not to the failure of a critical part, but to a simple cause such as inadequate tension on a limit switch spring causing arcing, or failure to provide a sufficient flow of oil to an apparently unimportant and lightly loaded bearing.

Design Review Even when the new design follows stereotyped lines, but especially when it must of necessity be an advance beyond existing practice, the designer is strongly advised to follow a deliberate plan. This should take the form of a design review, a method which has been proposed by reliability engineers.

It is no more than a detailed consideration of each section and each constituent part of the whole design, to contrast it with previous experience. Where there is no existing practice to draw upon, consideration must be given to the planning of tests to provide the knowledge which is necessary to ensure a sound design.

This review should be carried out along formal lines. In a well-organized design department there will be an accepted pattern, dividing the company's typical product into major sections. These should be listed in the usual manner, and in each section all but the small detail components should be itemized. There should be a column indicating which previous model, if any, most closely resembles the new design. Another column should indicate that the new design incorporates all the design changes found necessary in the existing product. In another one, all points of difference should be stated. Finally, it should be stated against each item whether or not it can be accepted with confidence in its proposed form, and what action is proposed to acquire the necessary confidence.

This may appear to be an obvious step to take. Unfortunately, it is carried out all too rarely. Often, insufficient time is the excuse. Sometimes, the designer presumes that he knows best and resents the design review as a reflection upon his ability. Yet where this painstaking and detailed check is carried out, it is a continual source of surprise to those concerned that so often new factors are brought to light.

It is always far cheaper to erase a pencil line from a drawing than it is to throw away a piece of hardware made to that drawing. Management should, therefore, be aware of the human problems involved, should ensure that the programme allows sufficient time for a proper review to be made, and should insist that it is in fact made.

To be effective, such a review is heavily dependent upon the availability of information regarding the way in which previous products have performed. This is an important matter to which adequate discussion will be devoted later. The tests which will be found necessary to fill gaps in available knowledge are the joint responsibility of designer and development engineer.

When the tests of new sub-systems and sub-assemblies have been concluded satisfactorily, the design of the complete product can be finalized.

Finalizing the Design The stages so far completed should result in a final scheme representing the designer's opinion of what he judges to be capable of meeting the customer's requirements. No design should be cleared for manufacture, however, until it has been confirmed that it can be produced at minimum cost.

Where the product is a 'one-off' there is no problem. When the product is intended to be produced in large quantities, a number of prototypes will be made for test. This number may vary from one prototype in the majority of cases, to two as in the case of the Concorde, upwards of a dozen for aero engines, a hundred or more for a new model of motor car. There will be considerable pressure from management to start these tests as soon as possible. In the majority of firms these prototype models will be made in experimental establishments, and all too often experience will be gained on machines which will not accurately represent the final design for quantity production (see page 151).

Nevertheless, at the earliest time practicable a thorough check should be made to ensure that the projected design will be economically producible. This requires close and effective collaboration with other departments in the organization. Since it is not unusual to find that a state of tension exists between these departments, it is a direct responsibility of management to see that a system exists which will ensure that there will

134

be close liaison between the engineers concerned with design, manufacturing, quality, value, purchasing, and servicing. The following list
summarizes the aims of this collaboration. It will serve as a check list,
enabling management to approach this hitherto sacrosanct area of
design with a degree of confidence.

*Objects of Liaison between Design and Production, Quality, Value
Engineering, and Purchasing Departments*
  (a) To make cost comparisons of alternative designs.
  (b) To reduce the number of components.
  (c) To use standards (domestic or trade or national) where possible.
  (d) To reduce the number of production operations and processes.
  (e) To maximize material utilization.
  (f) To use new materials and processes which offer an advantage.
  (g) To keep within existing manufacturing capabilities.
  (h) To ensure that critical components can be properly inspected.
  (i) To confirm that standards of acceptable quality are defined.
  (j) To check that all instructions and drawings are clear, unambiguous,
       and understandable by all concerned with their implementation.

When the design has reached an acceptable stage of feasibility, a
complete specification should be prepared. This should lay down the
requirements which must be met by products which may be numbered
in ones, tens, or hundreds of thousands. It should include a complete
set of arrangement drawings, showing to scale the construction of the
product. With complex products there may be hundreds of such drawings,
covering all the sub-sections and sub-assemblies of the complete product.

These drawings should be supplemented by a written specification,
indicating the precise dimensions of components which must meet
critical strength requirements. Vital working clearances should be specified, but these, and important dimensional tolerances, should be finalized
only after discussion with the manufacturing engineers, to confirm that
they can in fact be achieved without undue difficulty.

A list should be prepared of the materials which are to be used, with
the special properties, e.g. of strength or hardness, which are necessary
to meet the stress conditions. Critical areas of the design should be
indicated when necessary with a note that standards of quality will
have to be evolved in collaboration with the quality engineering department as experience is gained in production.

Special conditions of checking or test, e.g. of a welded structure,
should be stipulated. Where proprietary articles are to be incorporated
in the end-product, the required performance should be specified in full.

The foregoing amounts to no more than what is obviously necessary to enable those concerned to ensure that the end-product will in fact meet the design requirements. The sad truth is that only a minority of design departments recognize these responsibilities. All too often the interpretation of inadequate instructions is left to planning engineers, inspectors, buyers, and others.

It is always quicker and cheaper to get every job 'right first time'. Those who aver that there is not time to prepare the specifications can always find the time to put things right after they have gone wrong.

Management has a distinct duty to see that the design function is properly discharged.

The Detail Specification The next stage is concerned with the conversion of the design specification into a form which can be translated into working instructions by the production engineering department.

Practices vary widely, in different industries and in different firms. Much depends upon the size of the firm, the rate and scale of output, and the type of product. In small firms making one-off products which do not differ greatly from previous ones, it is not unusual for all the information to be recorded on the minimum of arrangement drawings, which carry dimensional sketches of detail parts. Sometimes existing drawings may be altered-up in blue pencil. In such cases, great dependence is placed upon the ability of the shop foremen and operatives to follow established practice, to decide for themselves the finish, the fillet radii, the heat treatment, and the bearing clearances required for the satisfactory functioning of the product. Indeed, within the past 25 years at least one large builder of steam turbines left it to the operative to turn a large shaft to the nominal diameters, and then to machine bearings and turbine disc bores to suit. Toleranced dimensions were not thought necessary, for 'real engineers'.

Today, it is better appreciated that an accurate record of what has been made is a valuable commodity. Even a one-off design may be so successful that a repeat order is won. In any event replacement parts might be required at some future date, and they must fit. In the case of large-scale production, it is quite usual for parts to be made in several different factories, and of course interchangeability is essential.

It is now the general practice to prepare a separate 'detail drawing' for each individual component. This drawing indicates the shape of the article, its dimensions, and the permissible variation in these dimensions, the material and its characteristics, details of finish, special conditions of inspection or test, and acceptable standards of quality. In the interests of consistency and to avoid overburdening the drawing with information,

136

it is preferred that information regarding materials, heat treatments, finishes, processes, inspection operations, and standards of acceptance should be conveyed by reference to standard documents, which should be listed on the drawing.

The preparation of these detail drawings, specifications, and standards is usually carried out in a separate detail drawing office. The draughtsmen may not have the creative ability required of the designer, but they do need to have a comprehensive knowledge of the practices and capabilities of the company's production shops. This must be combined with an understanding of the technical requirements of the design, so that the detailing office comprises a link between the design and manufacturing activities, which should result in satisfaction to both parties.

In many companies, the detail drawing office is given the responsibility of adjusting the design to the capabilities of manufacture. The extent to which this should be allowed will depend upon the degree of sophistication of the product.

Some American companies meet this situation by having a production design department which 'productionizes' the developed and proven design. While this results in end-products of a simplicity and cheapness which are the envy of their British competitors, there are disadvantages. There is the risk of loss of refinement of the design, while the designers proper tend to lose touch with modern developments in manufacturing processes. Where competition is intense, the quality of design may be the vital factor. Dain[1] has presented in detail the phases of the design cycle from market need to finished product (Fig. 10.1).

Making the Best of Specialization New materials and processes, new technical developments and requirements are appearing at a greater rate than ever before. A consequence is that it becomes increasingly difficult for the individual engineer/designer to keep abreast of the knowledge relevant to his work. At the same time technical education, in the United States as well as in Britain, is tending to become more narrowly specialized.

Formerly, the training of an engineer maintained a careful balance between theory and practice. This enabled him to take fuller advantage of the capabilities of the men in the shops, whose aggregate experience of things practical inevitably greatly exceeded his own. Today's emphasis upon the importance of a university degree, and the inability of most university teachers to communicate the intuition which is an essential part of engineering, are depriving industry of much of the art of the engineer.

Until the engineering profession learns from the medical profession that to turn a student into an engineer he must 'walk the wards of industry', an additional burden must be shouldered by management.

The disadvantages of narrow specialization must be minimized by insisting that specialists must work closely together in a common effort to evolve the best product of which the organization as a whole is capable. Management must meet the challenge posed by the problem of bringing together in amicable cooperation men and departments who, at the present time, are developing divergent attitudes.

In some companies, a useful step has been taken in the direction of closer integration of design and manufacturing interests. A small number of experienced production engineers have been stationed in the design department, to help the designers to evolve more easily producible schemes from the earliest stages of conception.

Of course this needs to be extended to include the quality, inspection, purchasing, and service activities as well. Not every firm has staff of the size to permit of interchange along these lines, but the need to achieve close collaboration is present all the time.

References

1. Dain, R. J. 'The responsibility of the designer in relation to quality and reliability-design as a systematic process.' Proc. Discussion on Quality and Reliability, Institution of Mechanical Engineers, London, 1963.
2. Haynes, W. A. 'Quality in warships.' Transactions Institution of Engineers and Shipbuilders in Scotland, January, 1966.

11

Proving the Design

No matter how thoroughly a designer may have applied his creative ability, his experience, and his technical knowledge to the task before him, management should never forget that he is human and, therefore, fallible. Often, he will be subjected to considerable pressure to complete his design quickly, by senior management not fully aware of the importance of the design phase. In consequence, he may be compelled to devote less time to some aspects of the total design than they deserve.

In the effort to evolve a design in advance of those of competing firms, the designer may be faced with problems outside his experience. His technical knowledge may not be quite adequate, yet he may be encouraged to take a chance—a leap, maybe ever so small, into the unknown. In addition, the product may require the preparation of hundreds or thousands, or more, of detail drawings. It would be an exceptional design department indeed which could guarantee that no mistakes would be made in such an extremely complex and often hurried situation.

Even where management has been persuaded of the importance of the design function, there may be a failure to realize that the designer who works alone, without the contrapuntal challenge of the development engineer, is deprived of an aid of great value.

It is unfortunate that up to the present time the development function has not found so vigorous a champion as has been the Feilden Report for design. Design is obviously the first essential step, but it needs to be supported, checked, and confirmed by test. 'There is no safe way of judging anything without experiment', wrote Henry Royce.

Development Justified
Failure to appreciate the importance of the proving phase is not confined to senior management. Some designers resent submitting their work to

139

independent check. Production engineers, already under considerable pressure to meet tight delivery dates, want to begin work as early as possible. The development engineer is under pressure from all sides to curtail his testing time. Examples come under notice daily of products which are failing to come up to expectations, and incurring heavy financial losses, because of some minor deficiency which would have been revealed by testing. In general, this testing would have cost only a fraction of the consequential losses.

The importance of proving the design is not confined to complex engineering products. Financial risks must be considered relative to the investment in the project. It must be realized that they can be as great with a relatively inexpensive item manufactured in large numbers as with a costly one-off installation whose purpose is to provide increased earnings by its efficient performance. A few examples will confirm the argument.

> A large retailer launched a new and attractive design of lady's nightgown. It was styled, planned, tooled, produced, and distributed to hundreds of shops. Only when the first customer went to bed in one was it discovered that inadequate armholes pinched her when she lay down.

> The same retailer had to recall thousands of garments because the zip fastener jammed in the open position. In the interests of increased productivity a new method of insertion of the fastener had been adopted, which did not provide a positive stop.

> Defects in motor cars attract greater publicity, especially in the conditions of 'total recall' now imposed by the American Motor Car Safety Act, 1966. The chafing of a flexible hydraulic brake hose, the risk of exhaust gas entering the passenger compartment, an insecure door lock, a steering column which does not comply with the crash requirements—any of these can and has entailed the withdrawal of millions of cars from service, for rectification.

These troubles have occurred with the products of firms enjoying a reputation for good design and reliability. They emphasize the importance of the development function. Sometimes a company may already have a sound policy of thorough testing. Even here it has been known for last minute changes, to correct a deficiency revealed by test, to be taken on trust. The changes have been thought to be insignificant, yet serious trouble has arisen from some cause which was not apparent, but which a test would have revealed.

> A zinc-based die-cast backing washer was substituted for a steel pressing, which was cutting into the cup-leather in a hydraulic

140

brake-system. As the zinc alloy aged it gave off gas which entered the brake fluid. Thousands of vehicles had to be campaigned to replace the washers and vent the system.[1]

Nylon bushings were attractive for sewing machines. They were cheap to make, wore well, and were quiet. Unfortunately the nylon was slightly hygroscopic, and in time the bushings distorted. This was an example of failure to consider all conditions of environment, and all the properties of the material. Research has since produced a stable nylon, but in addition the engineers have evolved a more comprehensive test programme.[2]

Of much greater and direct effect upon the individuals concerned was the failure of a new type of artificial hip-joint, which also used nylon. Implanted enthusiastically into hundreds of arthritic hips, they all had to be replaced. It transpired that the nylon was incapable of withstanding body fluids for a worthwhile period of time.

Failures of large capital equipment appear to be better publicized. They have precisely the same root cause as in the examples just described, a lack of sufficient testing to prove the design. This is almost invariably due to the lack of appreciation of its importance, and to the impatience of all concerned to bring the project to practical (albeit often ineffectual) realization.

In a land based steam turbine the absence of a programme of stiffness measurement in an entirely new design resulted in stresses being imposed in service which were many times greater than calculated. A shaft failed after a very short period of running. The commissioning of a power station was delayed by several months and heavy penalties were incurred by the contractor.

An expensive and potentially dangerous failure on a newly electrified railway was due to the fact that neither the British Railways Board nor the contractor deemed it necessary to test the locomotives under conditions approximating to those in service.

Examples such as the foregoing could be multiplied, in every country. That no country has a monopoly is shown by reports of trouble in the power supply industries in the United States, which parallel those being experienced concurrently in Britain. The largest American contractor in the business is reported to have lost $50m on one contract, due to a constructional failure.[3]

Confronted almost daily with such evidence, it remains a source of continual surprise that management does not take the initiative. Far more often than not the trouble is due to some minor component. The designer, knowing that he is not being allowed sufficient time, takes pains with the more novel or difficult problems. The less important details receive less attention, yet they are the commonest cause of trouble. As we have seen, and as we shall see again, a simple straightforward test on a single sample of product, continued until three or four different components had failed sequentially, would have been sufficient to reveal the major sources of weakness.

In such cases as have been described, is it fair to blame the engineers, or the management which has not allowed them to do their job?

It is the present concern to convey to management a better understanding of the purpose of development testing, and of methods which development engineers can be encouraged to adopt in order to keep the cost of their work to a minimum.

Research and Development

It is not impossible that some managements may have discouraged development activities in their organizations because they have been confused by popular terminology. Seldom is development discussed in isolation. It is all too often coupled with research, as R & D, as if the two were one and the same subject.

There are cases where research merges imperceptibly into development. There are many more where research, in the true sense of the word, is not necessary, but where development testing of the simplest practical type is vital. To couple this latter activity with 'research' is an almost certain way to encourage the senior managements of the firms concerned to take an antipathetic view. They have seen and heard of the large research establishments of large companies and of nationalized industries. They have heard and read the utterances of Government ministers, exhorting British industry to do more research to keep abreast of foreign competition. They feel instinctively, and often rightly, that this kind of research is not for them. But they fail to see the fundamental difference between research and development testing of the kind in which British engineers excel.

Development is taken here to signify the engineering, technical, and testing activities necessary to confirm the assumptions of the designer, to provide him with information to enable him to amend his design when necessary, to establish the viability of the product as a whole, its ability to perform as expected by the customer, and to assist in the continuing process of product improvement.

The Development Function

The development engineer can be seen to have a fivefold function:

(a) To obtain and provide the designer with information regarding the strengths and weaknesses of new processes, new materials, new concepts.
(b) To check the validity of essays by the designer into unknown fields, to confirm and to optimize the performance of new conceptual arrangements.
(c) To prove the final design of the entire product as being suitable for manufacture to meet the user's needs or expectations.
(d) To draw attention to unusual difficulties which will require special care by the production department.
(e) By providing a continuous flow of information based upon experience, to contribute to the improvement of current and future designs.

The emphasis which should be placed upon each of these areas of activity will differ with each type of product. The more extreme demands of functioning products which must meet high reliability requirements, such as motor cars and aircraft, will justify maximum effort in each of them.

With less vital products such as household equipment it is rare for a new model to differ from existing types in more than a few details. There will be fewer problems, and the main development activity may be confined to the identification of the one or two components likely to cause trouble in use.

Static or non-functioning products, such as clothes and furniture, will require effort mainly in the first category, little in the others, save perhaps for wear and/or laundering tests in the case of clothing and upholstery.

The five points listed above provide convenient headings for a more detailed account of what is involved in development testing, and, in a general way, of how the development engineer approaches his task.

Information for the Designer Every new design is, hopefully, an advance over previous ones. It may attempt to achieve something which has not been done before; or to do something more quickly, or more cheaply than before; or to give long life or more reliable service, or combinations of these desiderata. Whatever the aim, the designer will need information, about the capabilities of new materials, components, or processes.

Seldom will the information available from those who evolve the new materials and processes be adequate for the designer who has a specific project in mind. Unless they have been developed with this precise purpose in view, the data will almost inevitably be too general.

The information needed by the designer will often be provided by the development engineer. There is a somewhat hazy distinction here between research and development, but whatever work has to be done in the research laboratory, on materials or processes for instance, it will always be advisable to involve the development engineer as well. His task is to determine the capabilities, the strengths and weaknesses, not against a theoretical background, but in conditions simulating those of the intended use.

Often, the required information can be obtained quickly and cheaply. A simple method is to prepare a list of the available data, regarding the material, process, or proprietary article which the designer is considering using. This list should be set against the conditions of intended use, amplified by the development engineer from his experience of the way in which the products of his firm are actually used and abused at the customer's hands. Any doubts about the shortfall of capability for duty should be resolved by appropriate tests.

A good development engineer will be capable of judging by instinct what are the critical factors in a particular design. Simply dropping a piece of constructional material or a proprietary article on the floor may demonstrate its vulnerability to accidental damage. Examples have already been given showing how easy it is to overlook a time dependent condition which can lead to trouble (e.g. the nylon bushes, hip joints, and de-gassified zinc alloy already described). Unless the material experts can give positive assurance that the development engineer's fears are unfounded, the designer should use an alternative until the new material has been cleared.

Often simple 'lash-ups' will suffice to confirm the hopes of the designer.

> Sir Christopher Cockerell confirmed his hovercraft concept by means of a small model in balsa wood, which weighed only $4\frac{1}{4}$ ozs. It was capable of travelling over water at 13 knots. That was in 1955. Thirteen years later hovercraft weighing 150 tons were travelling at nearly 40 knots.

Such successful use of models demands engineering ability of a high order. Patents Offices are overloaded with inventions of prime movers which have worked on a model scale, but which have turned out to be quite impracticable when made in useful sizes. The inventors had failed to estimate the effects of size on heat capacity, heat transmission, rigidity, working clearances, and other factors.

> In 1610 an engineer putting forward proposals for water-driven machinery to pump water from coal-pits wrote: 'But small modles often fayle and soune prove defective when they cume to worcke upon heavye and continuall weightes in greater proportions. . . .'[4]

144

At the other extreme, it has been necessary to spend millions of pounds to bring the process for the production of a synthetic fibre to a realistic stage.

The designer can be helped considerably by being provided with more accurate and comprehensive data regarding the loads imposed in the actual condition of use of a product. Design rules established by experience often lead to 'overdesign'. Playing for safety, the resulting product is heavy and its performance is diminished. It is relatively easy with equipment which is available today to measure the frequency and intensity of loads applied during actual use.

> A producer of chain-saws measured the torque input when cutting different types of timber, the duration and frequency of use at the hands of the most vigorous lumberjack, the shock loads due to stones or nails buried in the trees. He was able to develop a design of much greater life and reliability than competing models. As a result he was able to raise his price while still providing his customer with better value.

That great engineer, the late Dr C. F. Kettering ('Ket') often said that the best work was done by those having inadequate facilities. This meant that people had to exercise their brains the more. Management should not, however, take this as an excuse for refusing to authorize the provision of facilities which may be essential. What should be realized is that lavish expenditure on test equipment is not a substitute for brains.

Component and Rig Testing The second phase of development activity is applicable to design projects which embody novel features. It is unusual for a design to be wholly new, and this allows work to proceed, ahead of the completion of the whole project, to establish the soundness of the designer's assumptions in those components and assemblies on which there is no previous experience.

A great deal of useful work can be carried out on rigs which enable proposed arrangements and performance to be confirmed. Thus, an electric circuit can be approved by making a 'bread-board' model, an arrangement of the detail components and the circuit, tacked on to a board, without regard to the problems of stowing it into a safe and compact container. The performance of a new design of turbine can be established first by tests on a model of the aerofoil, in a simple wind tunnel, and confirmed by building a complete stage of the turbine which resembles the final design only in respect of its aerodynamic characteristics.

New designs of wheels and suspensions, for vehicles and for aircraft, are investigated by building a single unit and subjecting it to drop tests and fatigue tests. The wheel is revolved against a large roller fitted with protuberances simulating bumps of varying intensity.

Each gear train, and the complete gearbox of a new design can be tested on a 'back-to-back' rig. One gear train drives another against an external load so applied that the only power input required is that necessary to overcome the power losses due to friction and churning of the oil.

It must be realized that in carrying out such tests there is always the risk that some vital factor might be omitted. It is part of the art of the development engineer to minimize these risks, and certainly to ensure that the last stage—the proving of the final design—is as representative as it can possibly be of the conditions of use and environment. The importance which is accorded to the application of the right conditions of environment has led to the formation, in Great Britain and in the United States, of societies to promote enthusiastically the study and the application of 'environmental engineering'.

As firms enter new markets it becomes increasingly important to study the new conditions likely to be met. When those markets are overseas, it is all too easy to assume that what is good enough for home use will be good enough elsewhere.

A manufacturer of a washing machine, which had been developed in collaboration with producers of washing powders to achieve good performance in Britain, encountered serious criticism from Swiss housewives.

It had not been appreciated that different conditions prevail in the two countries. The polluted air in Britain so dirties a man's shirt that it must be laundered daily. The clean air of Switzerland meant that laundering was necessary only every second or third day. This, combined with the better heated rooms resulted in the shirts becoming loaded with perspiration, with which the machines could not cope.

We have considered the responsibilities of the designer and the customer in defining conditions of use and environment. However inadequate these definitions may have been, it is up to the development engineer to do his best to visualize what is likely to happen to the product in the user's hands. Even when designer and user have tried conscientiously to define all foreseeable conditions, it can almost be guaranteed

146

that some enterprising individual, possibly imbued with a sincere desire to improve things, will think of some new way of doing something to the detriment of the product.

Testing the Prototype Product It is an important part of the development engineer's responsibility to draw up a programme of tests which will simulate as far as possible all the conditions likely to be encountered in service use. These will naturally take account of the intended periods and intensity of usage—of power and time under load, for example. They will include tests under the environmental conditions of climate. They will attempt to impose all the conditions of abuse which can be imagined.

In the usual case, it will be quite impracticable to carry out tests of the full duration for which the product has been designed. An ocean-going liner's engines may be expected to last for many years, or until the ship has become obsolete. The engines of an aircraft in airline use may run for 3000 hours a year, and have a total life of 30,000 to 60,000 hours, at least as regards many of their major components. Motor cars are guaranteed by some makers for one year or 12,000 miles, by others for two years or 24,000 miles, by a minority for five years or 50,000 miles. Military trucks are expected to run for 20,000 miles between overhauls, or 1000 hours; tanks and tracked vehicles for 4000 miles, or 400 hours. A guided missile must be capable of being stored for a considerable period of time, of functioning correctly when required, and of continuing to do so for the duration of its once-ever mission.

In other cases, a different kind of durability is required. Before a new textile fabric can be adopted, for men's shorts say, it must be established that it has good wearing properties, and that it can withstand many launderings with every kind of machine and detergent on the market.

> At least one large retailer carries out tests with his chosen fabrics to confirm that no new washing machine, no new washing powder, will diminish their life, or cause them to shrink.

In the electronics and nuclear fields, it is the usual practice to make statistical estimates of the likely reliability of the complete product, throughout its life. This is done by determining the probability of failure, measured as the *mean time between failures* (MTBF) of each component, and combining these by the *product rule* to obtain the reliability of the whole complex system.

> If there are n components in a system, each having reliability r, the reliability of the whole system will be $R = r^n$.

r will usually be less than unity (a highly reliable electronic component will have a 99·99 per cent chance of survival, so that $r = 0·9999$).

The more components there are in a system, the lower will be R. For example, a system having 100 components in series, each with a reliability of 99·5 per cent, would have an overall reliability of only 60 per cent.

Since the early guided missiles were extremely unreliable, and as they were complex devices having as many as 300,000 components, what Dr Robert Lusser called 'the appalling arithmetic of unreliability'[5] gained considerable credence.

An extensive literature emerged, and efforts were made to extend the statistical approach into the mechanical field. While anything which will lead to a detailed examination of a design must be beneficial, it was not realized that there are important differences between electronic and mechanical systems. Indeed, to take the statistical approach to extremes can be fallacious in any case.

Briefly, mechanical components are much less unreliable than are electronic ones. Hence they produce smaller numbers of failures, which are far less statistically significant. Mechanical systems do not usually involve large numbers of similar components, and they have a long background of tradition and experience. A mechanical designer who evolved a washing machine in which every component was unreliable would receive short shrift from his employer. We know from experience that a switch or a hinge may give trouble, a belt may slip, or an electrical connection may be defective. But the great majority of the components will give no trouble at all.

Fortunately, more balanced views are beginning to prevail, and management may feel reassured that the amount of testing required to establish reliability is usually much less than the statistical approach would suggest.

Grinsted and Spurr[6] of the Royal Aircraft Establishment support the author's experience.

> A more practicable and worthwhile approach towards improving reliability is through thorough development testing that is aimed directly at bringing weaknesses to light at the earliest possible stage.

The engineer's approach is based upon the knowledge that a test which does not lead to failure of at least one component tells him little or nothing about the potential capability of the product. Such a test reveals the weakest links and a mark of good engineering ability is the

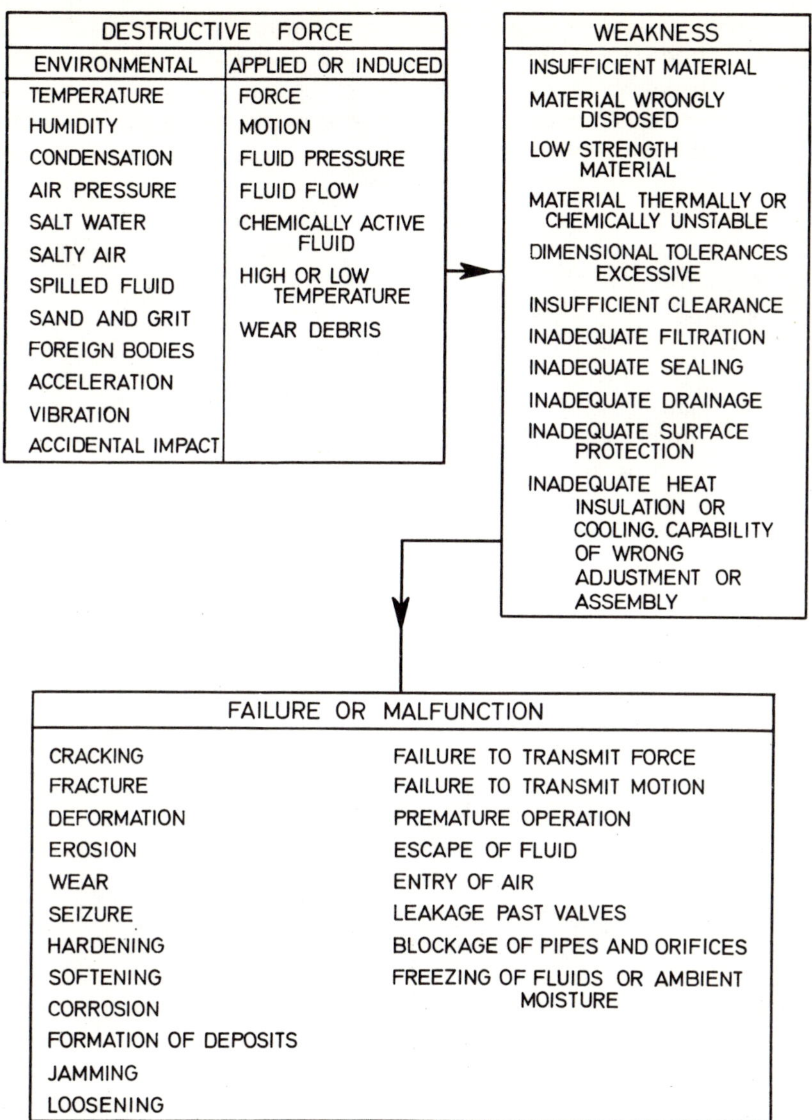

Fig. 11.1. Forces and Weaknesses in Mechanical Failure: an Aide-Memoire for the Development Engineer.[6]

extent to which it is possible to produce the kind of failure which is experienced in the field, by a simple test of short duration.

Engineering judgement is required to determine the safe amount of overload which can be applied, and a background of experience is a valuable aid to the prognosis of the most likely weaknesses in the design, so that a test can be devised to 'press where it will hurt most'.[7]

As in every other field of activity, development testing can be made more effective by adopting a methodical approach. A check list along the lines of that proposed for designers (page 132) will be useful. The response of the development engineer will, however, be different, insofar at being closer to the actual hardware and the conditions of manufacture and use, he will be able to bring to bear a complementary point of view which is one of his main contributions.

Above all, as will have been seen from examples already given, the development engineer should *never* omit to test every design change, however small or insignificant it may appear to be. Failure to observe this rule was the cause of serious troubles in the early Polaris submarines[8] and in many less well-recorded projects.

Figure 11.1, due to Grinsted and Spurr, provides a useful reminder of causes and effects which should be borne in mind.

When the project presents the development engineer with problems which come outside his experience, as when the design is so novel that there is no background of experience, a 'probe-testing' technique may be used. This has been proposed by Dorian Shainin[9] who has done a good deal to popularize simple applied strategies based upon sound statistical principles.

'Probe-testing' makes use of a pragmatic distribution evolved by Dr W. Weibull. By using a specially designed 'Weibull probability paper' it is possible to plan tests, and to plot the results, so as to obtain a maximum of information from less testing than purely statistical considerations would indicate. Whereas the statistician will call for tests on thirty products, by using the Weibull approach valid results can be obtained from seven tests. This method must, however, be used with discretion. There is always a tendency to ascribe to any new technique powers beyond its capability. The late J. H. Bompas-Smith has shown how the best results are achieved by blending the Weibull method with engineering judgement.[10]

In the experience of the author and his colleagues, it is often found that one or two tests will suffice to reveal a potential weakness in a mechanical device and to identify the cause of the weakness.

> Normally, one or two tests are required to confirm an analytical result such as stress analysis. Inter-action effects and confirmation of empirical results may require statistical analysis and designed experiments. . . .

150

It is important to remember that we are not evaluating an abstract set of numbers over which we have little or no control. We are estimating the strength of components about which we know a great deal.[11]

Hazards of Development Testing

As a general rule, the desire to see a newly designed product through its proving trials as quickly as possible leads to the taking of short cuts in the manufacture of the prototype. For example, parts which in series production will be machined from die-forgings may be machined from solid blocks of metal. Parts which will be cast in the final version may be simulated by fabrication. Methods of welding may be used which differ from those which will be adopted in production. Often, too, by a process of selection men of superior skill will have been recruited into the experimental shop, and they too will be impelled to take short cuts.

The excessive pressures which are sometimes imposed upon these people, to build the prototype as quickly as possible, leads to deviations from the drawings. Shortage of time precludes the proper recording of these deviations, so that the product as tested and approved may be different from the drawings and documents which should define it, and to which production units will be made.

The small number of prototype products makes it difficult to test parts covering the full range of variation which will occur, in dimensional tolerances for example, when large production quantities are involved.

Testing to Production Standards These practices simply build up trouble for the future. At the least harmful level, they fail to explore the full range of variation likely to be encountered in series production. At worst, they fail to test the product in the form in which the customer will receive it.

It is difficult, in the circumstances which usually prevail at this stage of product proving, to insist on meticulous observance of the experimental drawings, or to record all deviations. It is even more difficult to ensure that the prototypes shall accurately represent production models. The effort, however, can be well worthwhile. A notable example of the early achievement of reliability in service, with a complex guided missile which presented many novel problems, owed its success to strict observance of a rule of the company involved.

R & D Hardware must be to Production Standards

One of the easiest pitfalls into which development can fall is in the use of an 'over-intelligent' grade of works' labour to achieve early development working models.

Unless handled extremely carefully this is likely to result in the works manufacturing to inadequately defined drawings and instructions. It is very easy for development works staff to believe that they are doing their best if they make something, somehow, but this can lead to enormous subsequent difficulties when inadequately corrected drawings and instructions are released to production. While we still retain highly skilled personnel on our development works side, we use that skill to help us to apply a most rigorous system of works queries, so that inadequacies in drawing information are fed back to the Drawing Office for embodiment in the design documents. This is a dreadful and costly task, but unless squarely faced and dealt with during early development the problems will inevitably surface later, when larger numbers are being manufactured, and cause heavier penalties in cost and delay to the programme.[12]

Failure to follow this ideal practice means not only that prototype tests are carried out on products which do not accurately represent production models in respect of methods of manufacture, but that in addition, valuable information regarding changes of design which have been found to work satisfactorily, is not carried over into production. In recognition of the first difficulty, it is often a customer requirement that the acceptance tests shall be repeated on the first model or models off the production line. If the second situation has been allowed to arise, it is then too late to rectify errors of design, which are passed to the customer in the first batch of products. Increasingly it is becoming the practice in the automobile industry to rush through a batch of 50 or 100 cars made by production methods, ahead of the main batches, and to subject these to rigorous and intensive testing.

Variations in Production It is not as easy as might have been implied to explore, during the development phase, the effects of the variations in materials, components, processes, and final product which will inevitably exist in a production run of a large number of products. For one thing, the extent of this variation cannot be known until production quantities have become available. Hence, desirable as it may be to carry out proof testing on production-made units, the information which these tests can provide is limited. It will be a matter for consideration and decision by engineering and management to decide how much should be done at this stage, which in effect establishes a norm, and how much can be done by the quality engineering department, by methods which can be even more simply applied than the tests of the development engineer. This point is discussed more fully in chapter 9.

The Development Engineer

Asked in what sector of industrial effort they would prefer to work, the majority of young engineers will opt for development. The popular image of the engineer controlling powerful machinery to do his bidding, or conducting impressive tests on large structures owes something to television and the cinema. It is, however, far from accurate. The development engineer must be an engineer in the true sense of the word. His range of responsibilities is as great and as wide as that of the designer. He, therefore, needs to have the same level of technical knowledge and ability, save that instead of creativity, which is an important attribute of the designer, he needs a critical approach, and the ability to visualize external factors and their effects.

Since he must endeavour to seek out and reveal weaknesses in the design, many firms feel that he should operate separately from the designer. Whether he does or not depends a great deal upon the atmosphere prevailing in the organization. When the development activity is looked upon as contributory to the design effort, with the prime function of confirming the soundness of the design, the precise pattern of organization is of minor importance. When the designer looks upon suggestions from the development engineer as criticisms of his own ability, then it is important that the development department should be free to criticize.

Vested Interests The designer should be encouraged to have a vested interest in success, the development engineer a vested interest in failure. Psychologically, the designer is on his mettle to produce a product which will be satisfactory in every respect. The fewer the troubles which arise during development, manufacture, and use, the better a designer he is. This sometimes develops into a resentment of criticism from development, production, or customer, and this tendency must be rigorously discouraged.

At the same time, as we have seen, development testing contributes little towards establishing the capabilities of the product unless it reveals weaknesses. Hence, should a test not lead to failure of some part, then either the test has been wrongly planned, or by some miracle the designer has succeeded in emulating the Deacon's 'one-hoss shay', in which every component was of equal reliability and they all failed at the same time. In real life such things do not happen, and it is an indication of the development engineer's ability that he should be able to cause parts to fail in the same way that they would fail in service, even though his test perforce cannot simulate the service conditions accurately, either in method of loading or in total duration.

For these reasons, the development and design activities are probably best kept separate, coming together under the chief engineer in the small firm, or the project engineer or project manager in the more complex situation. Gunnar Brynge, however, has postulated that it should be possible to inculcate into the designers the frame of mind which welcomes rather than resents criticism and the deliberate breaking or failing of the products of their expertise.

It should be realized that the development engineer has a multi-faceted function. He audits the design by bringing to bear a special critical faculty. He identifies potential weaknesses in the design and assists in their elimination. He helps to bridge the gap between design and manufacture, by endeavouring to foresee manufacturing difficulties and the variables which will arise, assessing their effect upon the realization of the designer's intentions. He studies the historic performance of previous projects, applying the knowledge so gained to the further improvement of subsequent products and so acting in the dual role of agent of the user and protector of the designer's (and the company's) reputation.

Many examples of this engineering approach have been given by A. C. Lovesey.[13]

A particular difficulty facing the development engineer is that of communicating his views to non-engineers. Often, apparently insignificant indications of pending trouble can be seen after a short development test. The trained engineer recognizes these as pointing the way to serious failures after a longer period of use. He finds it extremely difficult, however, to convey his fears to the designer, and more especially to the customer's financial people, from whom he may be seeking sanction for a design change.

This difficulty is at its greatest in the case of equipment for the armed services. Until trouble has actually been experienced, it is not easy to convince the powers-that-be that a design change is necessary. Avoidable costs of great magnitude are incurred because heed is not paid to the opinion of the experienced development engineer.[14]

References

1. Martin-Hurst, W. F. F. 'The responsibility of development in relation to quality and reliability.' Proc. Discussion on Quality and Reliability, 1963. Institution of Mechanical Engineers, London.
2. Brynge, Gunnar. 'Design for Quality.' Vth Annual Conference, European Organisation for Quality Control, 1961, Turin.
3. *The Times* Business News, 24 December 1969.
4. Report on the Manuscripts of Lord Middleton, HMSO, page 174, 1911.

5. Lusser, Robert. 'The notorious unreliability of complex equipment.' September, 1956: 'Reliability through safety margins.' October, 1958. U.S. Army Ordnance Missile Command, Redstone Arsenal, Alabama, U.S.A.

6. Grinsted, F. and Spurr, H. G. 'The quest for mechanical reliability in aircraft.' Fourth Congress of the International Council of the Aeronautical Sciences. Paris, August, 1964.

7. Nixon, F. 'A planned approach from British industry.' Proc. Symposium on the Reliability of Service Equipment. March 1960, London. Ministry of Aviation and War Office.

8. Rickover, H. G. 'Quality—the never-ending challenge.' *Nuclear Engineering*, February, 1963.

9. Shainin, Dorian. 'Proving the Design in Quantity Production.' XIth EOQC Conference, 1967, London.

10. Bompas-Smith, J. H. 'The determination of distributions that describe the failures of mechanical components.' 1969 Annals of Assurance Science, 8th Reliability and Maintainability Conference, Denver, Colorado, July, 1969.

11. Coutinho, John de S. 'Whither reliability?' 1st AIAA Annual Meeting. Washington D.C., June, 1964.

12. Ingamells, E. H. 'The place of research and development in the achievement of reliability.' Vth Annual Conference, European Organization for Quality Control, 1961, Turin.

13. Lovesey, A. C. 'Gas turbine development—thirteen and a half years in commercial aircraft.' *J. Royal Aero. Soc.*, Vol. 68 No. 644, August, 1964.

14. Poulton, E. A. 'Special problems in the military field.' Colloquium on Aircraft Reliability in Service. *J. Royal Aero. Soc.*, Vol. 70 No. 663, March, 1966.

PART 4

Converting Concept into Reality

12

Manufacturing Good Product

The engineering department having laid down the specification of the product has thereby assumed the major share of the responsibility for its satisfactory performance. It hands over to the manufacturing department the responsibility for producing goods conforming with the specification. Conformance with specification involves much more than the 'quality' which the term defines. Unless the goods produced are in accordance with the designer's intentions, there can· be no knowledge of what works or does not work satisfactorily. There can, therefore, be no progress.

In addition, the manufacturing department must do everything it can to minimize the costs of manufacture, to reduce the losses due to quality failures, and to raise the overall productive efficiency. By this means, it will help to increase the value of the product to the customer, raising both the competitiveness and the profitability of the enterprise.

In three countries, Britain, the United States, and Japan, production engineers and managers have made a greater contribution to the achievement of product quality and reliability than is usually recognized by quality control and reliability engineers. This has been due largely to the way in which senior management in these countries has appreciated the value of Q & R.

In Britain, it was the Institution of Production Engineers and the British Productivity Council which took the lead, and the National Council for Quality and Reliability was the outcome of their joint interest.

We have seen how the work of the Bell Telephone System was not fully comprehended during the early years of the quality movement in the United States. During the past decade, however, under the pressures deriving from the enormous expenditure on defence and space research, the Department of Defense and NASA have done a great deal to educate American industry in the essentials of *systems management*.

In Japan, it was senior management, after hearing a talk by Dr Edwards Deming, which took the initiative leading to the great educational campaign in quality control principles.

Manufacturing, and Quality and Reliability

In the examination which was made earlier of the practical interpretation of product quality and reliability, it was seen that *quality* in the industrial sense is concerned mainly with the quality of conformance with the design of the product as delivered. Reliability is much more dependent upon the quality of the design than it is upon the quality of manufacture. This assumes of course that the quality of manufacture is maintained at the level of performance appropriate to the type of product being made.

Hence, those involved in manufacturing activities are concerned mainly with the quality of their products, rather than their reliability. The performance of these people is, however, greatly influenced by the reliability of the machine-tools and processing equipment which they use. The manufacturing units of companies, already realizing the importance of the reliability of their own products, have been quick to benefit. What is good and right for their customers must clearly be right for themselves, and they pass on to their suppliers of equipment the lessons learnt by their own firm.

The Involvement of Production Management

Important developments are taking place following upon the increasing professionalism of the production engineer. While to the pioneers of quality control must be given the credit for recognizing the importance of product quality, it is the production engineers and the managers who are putting the basic principles of Shewhart to practical effect.

Owing largely to the prominence which was given to the magnitude of quality costs before and during Quality and Reliability Year, the management of many British firms became interested. As a result, increasing numbers of production engineers and managers have helped towards the development of the following ideas. These are listed in logical order, rather than chronologically as they were appreciated.

It has been realized that:

(a) It is the production department, not quality control, which determines the actual quality of the product.

(b) The greater proportion of quality failures are management's responsibility.

(c) Employees and suppliers must be involved in the achievement of the overall objective.

(d) There must be effective collaboration with other technical departments.

(e) All activities affecting the quality of the product must be coordinated by detailed systematic planning.

The general managements of larger firms in technologically advanced industries have played the leading part in recognizing these factors. They must continue to spread the message among all their suppliers. More specifically, there is scope in many companies for increased efficiency in the performance of points (a) and (b) and for the development of a good system of paperwork. They must take the initiative in promoting more complete involvement and close cooperation between design, development, purchasing, manufacturing, and service departments.

Whose Responsibility?

Dr J. M. Juran has classified causes of scrap and rework as 'operator-controllable' and 'management-controllable'. He has assessed 'operator-controllable' error as being responsible for an overall average of 20 per cent of the total quality losses.

Operator-controllable error is seen to comprise:

(a) Accidental error, due to human fallibility.

(b) Wilful but well-intentioned error, due to deviations from instructions in the belief that a change will result in a better job or a saving of time, to the benefit of the company and the operative.

(c) Wilful, mischievous error, i.e. sabotage. This fortunately is so rare that it can usually be ignored.

It is thought that the quantitative effect of operator-error is much more variable than has been suggested by Juran's figure. It will depend upon the scope which is permitted to the operative to make a mistake, wilful or accidental. Thus, in a well-planned automatic or semi-automatic operation, provided that the operative has been properly trained and instructed, there is little that he can do wrong. There will always be the risk of an occasional aberrative lapse, but this can be detected quickly and corrective action can be taken before many parts have been made incorrectly. In such cases, it is unlikely that the operator will be responsible for more than 5 per cent at the most of the total scrap and rework.

In a private communication, J. Holmes has provided the following analysis, for a large well-planned automatic machine shop:

CAUSES OF SCRAP	PERCENTAGE
Unsatisfactory method	26·2
Faulty operation	1·3
Faulty setting	0·8
Poor jig design	41·4
Inadequate maintenance	6·5
Setting-up scrap	7·9
Machine-tool in need of repair	0·8
Faulty heat treatment	3·4
Defective material	2·5
Destructive testing	9·2

At the other extreme, where a good deal of dependence must be placed upon the skill of the operator, the figure might well approach the 20 per cent of Juran. Hand operations, processes not yet fully developed, depend upon the skilled and experienced operative, who might well be producing more scrap than a much less skilled man. The attention of a skilled and conscientious operative may wander should the operation cycle be slow. A man turning a large shaft for a steam turbine, or a large rotor, carries a heavy responsibility to maintain constant vigilance.

It is not easy to make truly objective judgements as to where the responsibility really lies.

The scrap in a machine-shop was running at 6·8 per cent. The cost of applications for concessions to accept deviating parts, and of rectification or salvage operations, brought the figure up to 10 per cent of the value of the total output of the shop.

The first analysis of the causes of this scrap suggested that 65 per cent was due to operator error. Further investigation, however, identified this 65 per cent as due to the following causes:

(a) Misreading of gauges.
(b) Lack of appreciation of the costs due to errors.
(c) Lack of familiarity with machines.
(d) Lack of attention to critical operations.
(e) Inadequate drawings and instruction sheets.
(f) Tools not available.
(g) Incorrect use of tools.
(h) Faulty tool maintenance.
(i) Inadequate machine capability for the accuracy required.

It was evident that the operative could have been doing his best, in circumstances which were largely outside his control. All of the causes were eliminated or greatly alleviated through action by the production engineering department. The action included the provision of direct reading gauges; a poster campaign, highlighting

162

the costs of spoilt work; training; improved provision of instructions and tools; control charts maintained by the operator.

Action similar to the foregoing is characteristic of Japanese as well as British factories. In Japan, care is taken to ensure that operatives are well-trained and instructed to carry out their work properly. Japanese management is extremely conscious of its responsibility for the quality of the output, and this is interpreted as a need to exercise leadership so as to involve every employee in the quality aim.

Improving the Situation

Following directly on the examples just given, it is interesting to see how the logic of the statistician can facilitate further corrective action. The control chart is based upon the assumption that the 'normal' pattern of variability in the characteristic properties or performance or results of action is due to the accumulation of causes which are accepted as inevitable to the process. They are classed as 'non-identifiable', i.e. not identifiable by acceptably economic action. The existence of abnormal causes is shown by departures from the normal pattern of behaviour, and these causes are identifiable and can be remedied.

Dr Edwards Deming sees *identifiable* causes of abnormal manufacturing performance as those due to action by the operator. The *normal* causes are those over which he has no control. They are common to each operative in a particular activity.

They should not, however, be accepted as inevitable and unavoidable. Being common to all operations, they are obviously the responsibility of management, which must identify them and take appropriate action.[1]

Although he has expressed himself very differently, Deming's concept is basically similar to Juran's 'managerial breakthrough'. This queries the validity of the usual managerial approach to quality problems. If the quality performance of an operation suddenly worsens, alarm bells are rung and all available resources are deployed until the situation is restored to 'normal', when the burst of activity collapses. But 'normal' might have been considered to be a performance of 93 per cent (i.e. 7 per cent of unacceptable work). Why, asks Juran, should management be satisfied with the current norm of performance? Why should not an attack be made upon the causes of the shortfall in performance which has been accepted as inevitable?

This is in fact the philosophy underlying most of the success stories, where quality costs have been reduced by as much as 3, 4, up to 10 per cent of turnover.

Deming lists the following common causes of variability in manufacturing performance, which are management's responsibility:

(a) Poor lighting.
(b) Humidity not suited to the process.
(c) Vibration.
(d) Poor instruction.
(e) Poor supervision.
(f) No specification or standards.
(g) Lack of interest of management in a programme for quality improvement.
(h) Poor food in the canteen.
(i) Inept management.
(j) Raw materials not suited to the requirements.
(k) Procedures not suited to the requirements.
(l) Machines not suited to the requirements.
(m) Mixing product from streams of production, each having small variability but different level.

It is interesting to note the resemblance between this list and items in the example on page 162.

Sometimes the identification and elimination of troubles such as these comes under the general title of 'quality action'. This is especially the case where top management has taken the initiative, and views the situation broadly. It has no conception of, nor would it tolerate antagonisms and jealousies between departments. Management ensures that 'quality action' is the concern of every department involved.

In other cases, even though the quality department may have identified the causes of trouble, it is usually powerless to do anything about them. It is in these more common cases that the interest of the production department is so welcome.

> In one company, the production engineer has taken the initiative. The decision has been made that managers of manufacturing units should be responsible for their total performance, for quality as well as quantity. Their profit plan includes, among other things, a budgeted annual reduction in scrap and rework costs.

By making production management responsible for quality performance from the outset, the task of implementing schemes of quality control and quality improvement is made much easier. Shop floor personnel are much more likely to be convinced by their own kind, whom they know to be responsible for them and their performance, than by a separate quality department. No matter how well management may have improved cooperation between departments, it will take time, in old-

164

established organizations, for all production operatives to look upon quality people as their friends.

Production management, too, is in the best position to assess its own situation. Very often firms with a large total output do not have long runs of a given type of product. A Ford plant in Africa, for example, produces 84 different types of vehicle. Effective performance in such cases requires a better than average understanding of the principles of quality control.

Reference

1. Deming, W. Edwards. 'Statistical methods as an audit of management in industry.' Annual Conference, European Organization for Quality Control, Madrid, 1968.

13
Buying Out

There are few industrial organizations which do not find themselves compelled to place great dependence upon other companies for the supply of materials, detail components, functioning devices, or services. In the case of large companies, such as motor car manufacturers, as much as 70 per cent of the finished product, by value, will have been bought-in. Many other firms obtain from 50 to 60 per cent from outside sources while some, for example a large designer/retailer like Marks and Spencer Ltd, or a defence department like the Royal Air Force, obtain *all* of their requirements from other firms.

One of the most important developments of recent years has been the recognition by management of the *interdependence* of supplier and customer. A prime supplier cannot succeed without the support and collaboration of those other suppliers from whom he obtains essential material. By the same token, if that prime supplier is not himself successful with his own customers, then his suppliers will eventually find themselves without a market. A leading firm will obtain essential commodites from a large number of outside sources, and these sources must become involved in the end-purpose of the primary firm. This is no more than a logical extension of the efforts of the main firm's management to achieve total involvement of the people and activities within its own organization. Though the external task may be much more difficult it is just as important, indeed often much more so. This involvement is an essential factor in the success of every project and it is one which deserves the serious attention of management.

Many people in Britain are familiar with the way in which Marks and Spencer have involved their suppliers in the plan to satisfy millions of retail customers. Basically similar in principle, but many times greater

166

in scale and complexity, the Apollo moon-landing programme deserves special mention. It represents the culmination of a sustained effort to identify, define, and promulgate the principles, practices, and responsibilities of two-party business. This effort was started by the US Department of Defense in 1955 and was adapted by the National Aeronautics Space Administration to its space programmes. Fortunately, this work has been well-documented and the information which is available is of great value to all managers, purchasing officers, and agencies, and to quality managers who wish to achieve the total involvement of all their suppliers.

In effective procurement systems, a common approach can be identified. The purchasing officer or agency, preferably in collaboration with the production engineer and the quality manager, will ensure:

(a) That the supplier knows what is wanted of him.
(b) That he is capable of doing what is required.
(c) That he understands his responsibilities, and undertakes to meet them.
(d) That he can be seen to have done so.

The formal, planned approach involves the preparation of an adequate specification, a survey of the intended supplier's facilities and abilities, discussions culminating in a written agreement which becomes the basis of a legal contract, and a check of the supplier's performance, which for economy and efficiency will be geared to the level of that performance. The better the supplier meets the customer's needs, the less surveillance and checking will be required.

Leading firms and Government agencies have evolved systems which are designed to ensure that they, the customers, receive satisfaction. At the same time, the suppliers themselves benefit greatly from the tonic effect of the applied discipline.

> The managing director of a British firm sub-contracting to larger ones described a quality survey as 'free consultancy'. Large supplying firms had at first resented being surveyed by their customers. Realizing the value to themselves of their own supplier-approval schemes, they had come to recognize the rights of their customers, whose surveys provided valuable independent checks.

Indeed, any company not already approved as a supplier to one of the leading firms in the country would find it well worthwhile seeking approval. They might at first find it hard to accept a critical review by their customers, but in the end they would surely benefit.

The Responsibilities of the Purchasing Department

The chief buyer of a company carries a weight of responsibility which is too seldom realized. He must spend large sums of money, often amounting to more than half of the company's turnover. With this money he must procure supplies of the right quality, in the right quantity, and at the right time to enable the firm's products to be manufactured to specified standards and delivery dates. With such a large share of basic costs within his control, he must ensure that he obtains the supplies at the lowest price practicable and compatible with the qualities required. To this end, he must conduct the affairs of his own department efficiently, and see to it that his suppliers take advantage of every practice and technique which will enable them to achieve maximum efficiency.

This is not the place for the discussion of details of the organization of the purchasing department, regarding which there is an adequate literature. What we are concerned with here are those functions and responsibilities which have a direct bearing upon the quality and reliability of the product.

These responsibilities should be laid down by senior management. Typically, the purchasing department should be charged with the responsibilities of:

(a) Familiarizing itself with the technical characteristics of the company's products.

(b) Studying and understanding the technical characteristics which are requirements of the materials and components procured from outside sources.

(c) Selecting only those suppliers who are considered to be capable of meeting the required standards of quality, delivery, and cost.

(d) Cooperating with quality and production engineering departments to confirm that the requisite understanding, organization, and technical ability does in fact exist or can be developed.

(e) Ensuring that the suppliers are provided with all necessary information.

(f) Cooperating with the quality department to ensure that the ability to produce parts to the required standard is confirmed by the acceptance of first-off samples.

(g) Obtaining deliveries of the requisite quality and quantity, at the times required.

(h) Holding discussions with the supplier to ensure that they fully understand their responsibilities.

(i) Drawing up terms of business, which will define
    (i) the responsibilities accepted by the customer,

(ii) everything which the supplier is required to do to meet his obligations to the customer.

(j) Incorporating such of these requirements as are legally binding into a formal contract between the two parties.

It is worth repeating that purchasing is essentially a two-party business. The joint and several responsibilities of supplier and customer have been discussed earlier. It is a primary concern of the buyer to ensure that each party is fully aware of what is expected of him.

For far too long the supplier–customer relationship has been bedevilled by lack of appreciation of the supplier's responsibilities. This has been due to lack of an attempt by the customer to involve his supplier in the total enterprise. Everyone concerned with industrial output will have experienced the cynical response to complaints of deliveries of poor quality goods. Dissatisfaction because of late deliveries is even more widespread. Root causes are the failure of the supplier to appreciate the effects of non-available or non-usable material upon a tightly planned production schedule, while the knowledge that he is in a seller's market acts as a disincentive to further effort. Parenthetically, the general excess of demand over supply which exists in Britain is in marked contrast to the situation in the United States, where it is usual to find several suppliers competing for one firm's business. This provides a powerful disciplinary incentive which is unfortunately lacking in large sectors of British industry.

The Responsibilities of the Supplier

Purchasing officers confronted with such difficulties as prevail in Britain may be helped by the Trade Descriptions Act 1968. This Act makes it an offence to sell or to offer for sale goods to which a false description is applied. When the goods do not comply with the description—the specification—the description is a false one, and an offence is committed.

The Act includes in the definition of a trade description all those things with which the engineering industry is familiar—quantity size or gauge, composition, performance, detail drawings, processes, quality standards. It applies particularly, therefore, to goods of proprietary type, where the supplier is responsible for the design of, for example, standardized components for motor cars, or items having wider applications, such as nuts and bolts.

In large scale production, we know all too well that with the best of intentions there will be the occasional defective product. A defence against a prosecution under the Act could be that all reasonable precautions had been taken and due diligence had been observed to avoid committing an offence. It would probably need to be shown that there existed an effective system of specification and instruction, and that every individual involved was aware of what he had to do, and had been provided with the means of doing it. Presumably, it could be an effective defence to demonstrate that non-conformance of product with description was due, in a particular instance, to a 'random cause'.

Since litigation is best avoided, it will pay suppliers to take every care to ensure that they have followed all the precepts and principles advocated in previous chapters.

The Act will be less likely to apply when a supplier is working to the instructions of his customer, since there will be no 'trade description' by the supplier unless this be in relation to the services which he provides. Unless all the details of the contract between customer and supplier are clearly recorded and understood by both parties, there will exist that grey indeterminate area which is so common in supplier/customer relationships.

Nevertheless, by imposing a disciplinary pressure the Act cannot but be beneficial as encouraging all supplying firms to take a more careful look at the descriptions and specifications of their products, and at the way in which their goods match up to those descriptions and specifications.

The Contract

The supplier/customer relationship can be optimized by the purchasing officer insisting that a legally-binding contract in clear and unequivocal terms be entered into between the parties. For it to be acceptable, the contract must make clear the duties and responsibilities of the customer as well as the supplier. Apart from the contract itself, it will help to establish mutual confidence and good relations if the customer makes a formal declaration of his recognition and acceptance of his responsibilities.

This declaration should include:

(a) A statement of the fact that the two parties are interdependent, and that the success of each depends upon the other.
(b) The announcement that to this end a more formal approach is being made, to ensure
    (i) that the supplier is clearly aware of his responsibilities.
    (ii) that the supplier and the customer are in fact capable of doing what is asked of them.
    (iii) that each party will help the other on all technical problems.
(c) An honest statement of the business prospects of the project for which supplies are needed—the immediate and long-term forecasts, with genuine statements of required delivery dates.

(d) An undertaking to give ample warning of any changes to the programme.

(e) The wish to see the supplier benefit as the customer's fortunes improve.

In turn, the customer is entitled to expect full cooperation from the supplier, who should be similarly honest and helpful.

A formal legally-binding contract has the important advantage that it emphasizes the seriousness with which the customer regards his business, and ensures that the supplier takes careful note of what he is undertaking to do.

For the achievement of the requisite quality and reliability of product as economically as possible, it is important to include in the contract items which have not always been considered worthy of mention.

The technical basis of the contract should be the customer's product specification, process specifications, the specification of quality assurance requirements, and quality standards.

It is assumed that due care will have been devoted to the compilation of these documents along lines already discussed. The contract should, however, place special emphasis upon the supplier's responsibility not to depart from the conditions laid down in these specifications, without the prior written agreement of the named representative of the customer.

It needs to be pointed out that no change shall be made in the organization, management, or methods of the departments concerned with the quality of the product (which include production tooling and processing, manufacturing, quality control) without written agreement. This is necessary in order to underline the importance of the quality system survey, and to help people to realize that what is approved is the system and the individuals functioning at the time of the survey. Unfortunately, under the Approved Firms' Schemes of some Government departments, there developed a tendency to think that 'once approved, always approved', and this has diminished the regard paid to approval schemes in general.

Supplier Quality Assurance

In 1966, a practical *Guide to Supplier Quality Assurance in Engineering* was published by the Institution of Production Engineers. It was inspired by H. W. Mander, lately of Automotive Products Ltd, past Chairman of NCQR, and it is based on first hand experience of the difficulties encountered in persuading some thousands of suppliers to adopt the principles of quality control. It is a useful complement to the official 'Requirements' described earlier, especially in its descriptions of how to cope with the tasks of carrying out surveys of suppliers' organization, systems, and facilities, and of inspection of goods received from the suppliers.

The scale of the effort required to indoctrinate a company's suppliers, and to conduct surveys of their quality systems, is less than might be imagined.

Examples of the effort required for quality system surveys are given below.

FIRM	NUMBER OF SUPPLYING FIRMS	NUMBER OF QC SURVEILLANCE STAFF	FIRMS PER OFFICER
1	1000	4	250
2	600	3	200
3	300	2	150
4	715	4	180
5	1200 (300 critical)	3	400
6	3400	10	340

These figures must be taken as a rough guide only. The amount of surveillance required will vary according to the importance of the product, and the extent to which the supplier has implemented the quality assurance requirements.

In addition to the permanent staff (sometimes called 'quality audit staff'), the quality survey team is usually free to call upon the services of experts in metrology, inspection methods, non-destructive testing, production planning and so on, to make brief visits to suppliers to render help and advice. This additional effort, however, does not sensibly alter the ratios shown above.

When a company is supplier of similar goods to a number of different firms, however, a confused situation can arise. Each customer will have his own 'quality programme requirements', which will have been based upon one or other of the two fundamentally different types of Government sponsored schemes (American, British Navy Department, *et al.* on the one hand, and British Ministry of Technology on the other). Each customer will wish to carry out his own survey and surveillance. In consequence, some supplier firms have suggested that their customers should be prepared to accept the approval of an independent organization.

Desirable as this may seem at first sight, it is not entirely feasible for two overwhelming reasons:

(a) Each customer has different requirements. Even though the same product may be supplied to different customers, each one will use it in a different way, and different standards of quality will

172

probably be required. These must be stipulated by each customer, in the light of his needs and of the duty which may be imposed upon the products.

(b) The customer must take full responsibility for the satisfactory performance of the end-product, including the bought-out components. This can only be done by direct, face-to-face two-party business. Dependence upon a third party leads inevitably to the abdication of responsibility, and eventually to a lowering of quality performance. What has been recognized as a fundamental principle by the American Department of Defense, when applied by numbers of large firms, must inevitably mean that suppliers will have to satisfy their different customers individually.

This is not to say that no effort should be made to follow a uniform procedure. Since the NATO documents AQAP-1 and AQAP-2 are being adopted by contractors in the NATO countries who wish to supply the NATO forces, there is a great deal to be said for adopting these, or at any rate the greater portion of them, at least as a basis.

At the same time, it will be in the interests of firms in a particular industry to agree on such parts of their requirements as are common. They should at least make the effort to agree upon a standard pattern of *pro forma*, so reducing the amount of paperwork which the supplier is required to produce.

Vendor Rating

The purchasing officer will be anxious to obtain value from his suppliers by encouraging maximum performance through minimum scrap, minimum need for acceptance inspection, timely deliveries, minimum avoidable costs. One way of achieving this is by establishing a 'comparative rating', and encouraging those suppliers with a poorer performance to make the effort to equal the better ones.

A variety of rating methods exists. In some, the desiderata of

Quality, as indicated by percentage of unacceptable parts received.
Delivery.
Quality of packaging.
Price, and so on,

are weighted, according to the needs of the purchaser, to produce a quality index figure. This has the drawback, however, that it can lead to 'trading-off' quality for speedier delivery, for example.

A better method is to rate vendor or supplier performance by confining the measure to the quality of incoming goods. Still better, and this is only possible where acceptance inspection is by sampling, is to establish a rating figure as the

$$\frac{\text{Desired cost of inspection}}{\text{Actual cost of inspection}}$$

which is found necessary by the customer.

> An advanced example of vendor rating is applied by one British company. Suppliers are categorized into A, B, and C groups according to the stringency of the sampling inspection which it has been found necessary to apply. If a delivery from an A firm fails to be accepted, and falls to B level, the acceptance sampling is raised to B standard until the position has been restored. To cover the increased costs of inspection by the customer, payments to the supplier are reduced by $2\frac{1}{2}$ per cent. If an A firm falls to C rating, the deduction is 5 per cent.

This scheme has been accepted on a legally-binding contractual basis by hundreds of the firm's suppliers. A scheme such as this, however, can only be adopted when a stable situation exists, and a satisfactory relationship, with complete mutual understanding between customer and supplier, has been established.

A Guide for Management

Management's position *vis à vis* its specialist staff is always strengthened when it is able to cross examine them and check their activities. The following check list will be as useful to senior management as to the executives in charge of all aspects concerned with procurement from outside sources:

A. *Check List for Customer's Purchasing Officer*
  1. Are the following aware of the intention to sub-contract?
     a Works manager.
     b Production engineering manager.
     c Production manager.
     d Quality engineer.
     e Chief inspector.
     f Controller.
  2. Are any of the following necessary? If so, do they exist?
     a Product specification.
     b Quality standards.
     c Process specifications.
     d Manufacturing technique specifications.
     e Acceptance inspection plan.

174

3. Do works manager, production planning engineer, quality engineer consider potential supplier capable of meeting the requirements? This should be confirmed in respect of:
   a Manufacturing ability.
   b Manufacturing equipment.
   c Knowledge of special requirements, e.g. Government specifications.
   d Inspection equipment.
   e Inspection ability.
   f Detailed internal organization and existence of necessary systems.
   g Do the necessary Government approvals exist?

B. *Check List for Supplier's General Manager*
   1. Are the following aware of the intention to accept a sub-contract order?
      a Works manager.
      b Production engineering manager.
      c Production manager.
      d Quality engineer.
      e Chief inspector.
      f Controller.
   2. Have they checked all the special requirements of the order?
   3. Are they aware of methods and techniques developed by the customer?
   4. Have they the necessary equipment and knowledge?
   5. Are they confident that they can meet the requirements?
   6. Do they accept the responsibility for delivering to specification and price?

Incentive Contracts

The idea of offering to contractors an increased profit for improved performance is recognition of the interdependence of supplier and customer. It has its widest application in defence contracts, where by reason of the novelty of the design concepts it is not always possible to evolve rigid specifications of requirements, and the contractor is himself often in a position to make a valuable contribution. It is recognized that profit is a powerful means of encouraging participation in a joint project.

Incentive contracting has received greatest attention in the US Department of Defense, and in our own Ministry of Defence (Navy Department) where as long as 300 years ago Samuel Pepys, then Secretary of the Navy, adopted competitive tendering for ships, with penalties for lateness and bonuses for early delivery. The basic purpose of an incentive contract is, according to Captain B. L. Lubelsky, USN (ret.),

> To assure attainment of the overall procurement goals for object-
> ives. In other words, the contractor is given an incentive to control
> costs, improve deliveries, eliminate over-design, improve quality
> or to increase or demonstrate System Reliability. . . . Quality,
> Reliability and hardware performance requirements must be
> established as *dominant* elements in the incentive earning
> potential. . . .[1]

Profit The profit motive has been clearly recognized by the American
Armed Service Procurement Regulation (ASPR) 3-402 (15 August
1963):

> Profit, generally, is the basic motive of business enterprise. Both
> the Government and its defense contractors should be concerned
> with harnessing this motive to work for the truly *effective* and
> economical contract performance required in the interest of
> national defense. Success in harnessing the profit motive begins
> with the negotiation of *sound performance goals and standards*.
> This objective is met if the contractor either benefits or loses in
> relation to achieving or failing to achieve *realistic* targets.

Cost Incentive In the simplest case, the incentive is offered to the contractor for achieving a desired quality and delivery target at less than the
estimated cost. This is the *cost incentive*. As a general rule, the benefit
of reduced cost is shared by Government and contractor in the ratio
80:20 or 75:25. On the other hand, the contractor suffers a penalty of
20 or 25 per cent of the increase when the cost is greater than agreed.
The norm allows the contractor a 10 per cent profit, and upper and lower
limits are fixed at 15 per cent profit for best performance and 5 per cent
for less than desired performance.[2] (See Fig. 13.1.) This is the *fixed
price incentive contract*. Alternatively, a contract may be based upon the
cost of manufacture and development plus a fee which has an incentive
content.

Multiple Incentives More recently contracts are being placed which
include multiple incentives, awards being given for reliability and
quality achievement, combined with cost and delivery, which are
weighted according to the importance accorded to them in each individual case.

Value Engineering Concurrently, Government contractors are encouraged to reduce the prime costs of products, by applying the principles of
value engineering. Owing to the directive of Robert McNamara, when
Defense Secretary, that everything possible should be done to contain

176

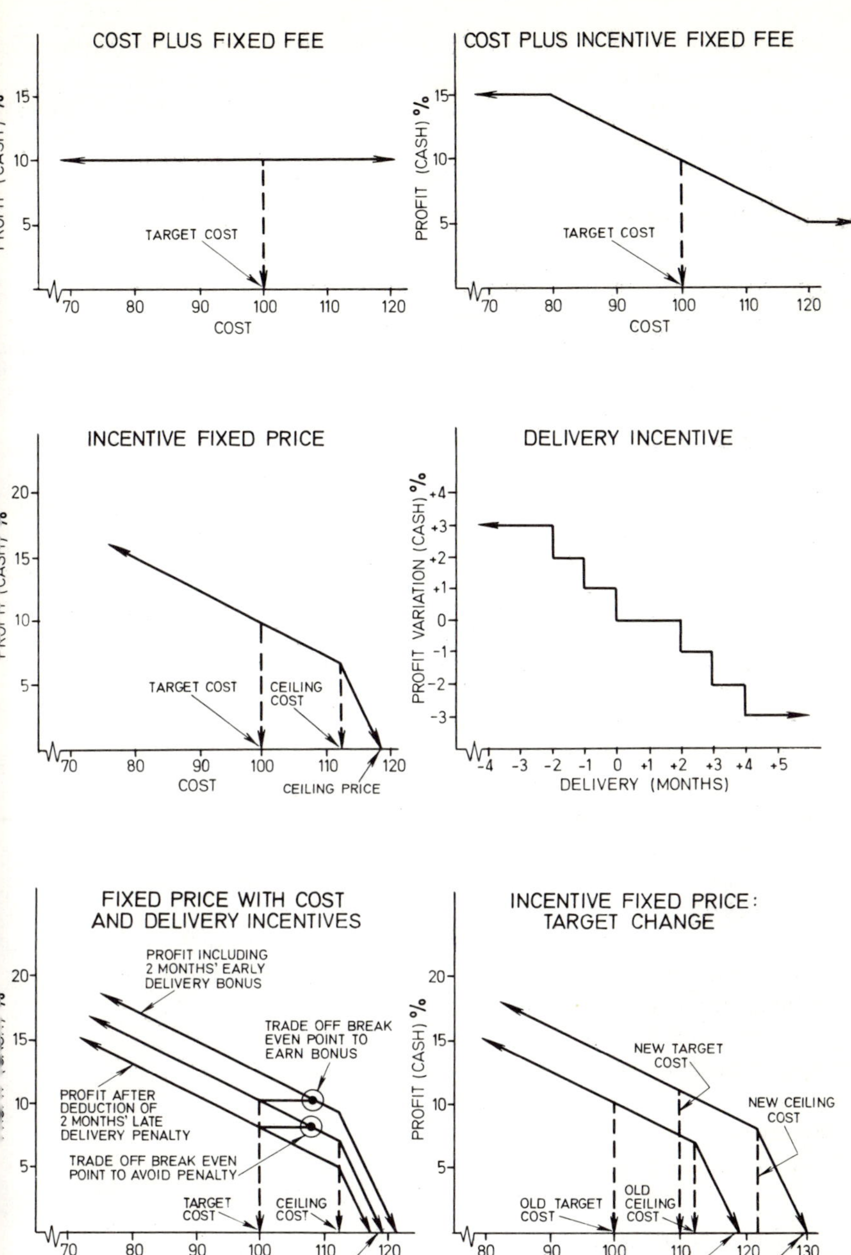

Fig. 13.1. Forms of Incentive Contract.[2]

and to minimize the country's defense costs, it is in the United States that we find the best-organized value engineering programme. In this, Government and contractor share the savings according to a predetermined ratio.

> The Goodyear Tyre and Rubber Company effected a reduction in cost of collapsible fuel tanks amounting to $752,000, by design changes. The company was allocated $205,000 as its share of the saving.[3]

Leading British firms have set up value engineering or value analysis departments. At first, there was a tendency to devote maximum attention to the cheapening of existing designs. It was soon realized that once a product has been designed, tested, and tooled for production a considerable amount of money has already been spent, and the hoped-for savings cannot always be realized. There were instances, too, where reliability was sacrificed. Today, it is realized that maximum benefit is obtained by close cooperation between designer and production engineer, from the earliest stages of the design. This ensures that the basic design offers the best chance of achieving maximum overall economy. This approach suffers from the unimportant drawback that because an alternative concept has not been finalized, it is not possible to quantify the saving.

Total Cost Incentives A recent development is to look upon a project in its entirety. Its value to the customer is measured by its estimated total cost throughout its whole life. To estimate this 'life cycle cost', or 'cost of ownership', requires sophisticated engineering of a high order, making use of design review and reliability engineering to forecast the probable life of the components which comprise the product. The effort to estimate total life cost is being made by the US Department of Defense and its contractors even with complex electronic equipment and nuclear submarines. The difficulty of this estimating, and the risk of error, may be appreciated by realizing that the total life cost of advanced military equipment may vary between 3 and 200 times the prime cost.

References

1. Lubelsky, B. L. 'Models for reliability and quality incentive contracting.' Xth Annual European Organization for Quality Control Conference, Stockholm, 1966.
2. Hedger, E. F. 'Incentive contracting.' Services/Industry Conference on the Reliability of Service Equipment. Institution of Mechanical Engineers, London, 1968.
3. 'VE nets firm $205,000.' *Defense Management Journal*, Vol. V, Issue 2, Washington D.C., Spring 1969.

14

Assuring the Quality of the Product

Customer satisfaction requires that every unit of product received from the supplier shall be of the specified quality.

How this shall be accomplished and *who* shall be held responsible have long been major topics for discussion among quality control men. The unfortunate result has all too often been management in confusion and quality control in disrepute.

The attempt will be made to put quality control in its proper place in the organization as a whole. It will be shown that despite the views of some proponents of quality control, inspection is an important and essential activity. It has already been made clear that quality control men, by themselves, do not assure the quality of a product, but they can and do make an important contribution. Of greater importance is the systematic managerial approach which is implied by 'quality assurance' as defined below.

For management to realize what it must do to set its house in order and to plan the system of communications, instructions, and actions which are necessary, it must be clear about the true functions and responsibilities of inspection and quality control.

The confusion which has hindered management's understanding of these activities is not confined to management alone. The experts themselves are not always in agreement. Nevertheless, John J. Riordan, who has done much to clarify the situation, holds management largely responsible. Over-emphasis upon one or other aspect of a new subject, he says, leads to the emergence of cults.

> Cults are assemblages of well-intentioned enthusiasts who strongly —sometimes fanatically—identify themselves with particular specialties (e.g., reliability, zero defects, systems engineering). To the extent that these cults both arise and persist, they are

179

symptoms of managerial oversight. Management simply failed to anticipate, confront, or resolve new and sticky problems.[1] Their failure to achieve recognition by management only serves to intensify the fervour of the cultists, and their claims become exaggerated.

Before proceeding to discuss what must be done to identify the needs of an organization in respect of product quality assurance, it is important that we should have some idea of what people in the quality field are talking about.

The Verbiage of Q & R

A cult does not become a cult unless there is a word to describe its specialty. Unfortunately, language is inadequate to convey in a word or two shades of meaning which vary according to one's own interpretation of them. Hence, there is a tendency to adopt words already in use, and to try to alter their meaning. Praiseworthy attempts have been made to avoid the ensuing confusion by preparing glossaries. These purport to set down definitions which have been agreed by groups of people.

These have met with limited success because, in a rapidly developing subject such as Q & R, it is inevitable that some people, and some countries, should move ahead more quickly than others. So, confusion and disagreement persist.

At the time of writing, the definitions of terms most likely to be clearest to the mind of management are felt to be those being revised by the American Society for Quality Control (1969–70). They are reproduced below.

For the benefit of firms doing business in Europe, there are shown, side by side with these, the definitions agreed by the countries subscribing to the European Organization for Quality Control. These may help management to understand the reasons why, in some countries, they may meet a rigid attitude of mind having its roots in narrowly professional quality control. This Glossary is valuable too because it lists key words in twelve different languages.

The American definitions show an increasing awareness of the management point of view. Indeed, it is gratifying to be able to recognize in the new definitions aspects which manifestly derive from the work of British writers.

DEFINITIONS OF TERMS

AMERICAN[2]	EUROPEAN[3]
Inspection	
The process of measuring, examining, testing, gauging, or otherwise, comparing one or more units of product with the applicable requirements.	Determination of the quality of materials, components, assemblies, or finished products, for purposes of process control, product control, quality audit, or fault diagnoses.

180

Quality
The totality of features and characteristics of a product or service that bear on its ability to satisfy a given need.

Quality Assurance
A system of activities whose purpose is to provide assurance that the overall quality control job is in fact being done effectively. The system involves a continuing evaluation of the adequacy and effectiveness of the overall quality control program (see Quality Control) with a view to having corrective measures initiated where necessary. For a specific product or service, this involves specifications, audits, and the evaluation of the quality factors that affect the specification, production, inspection, and use of the product or service.

Quality Audit
(Not included)

Quality Control
The overall system of activities whose purpose is to provide a quality product or service that meets the needs of users; also the use of such a system. The aim of quality control is to provide quality that is satisfactory, adequate, dependable, and economic. The overall system involves integrating the quality aspects of several related steps including: the proper *specification* of what is wanted; *production* to meet the full intent of the specification; *inspection* to determine whether the resulting product or service is in accord with the specification; and *review of usage* to provide for revision of specification.

The term 'quality control' is often applied to specific phases in the overall system of activities, for example 'process quality control'.

The quality of a commodity is the degree to which it meets the requirements of the customer. With manufactured products, quality is a combination of quality of design and quality of manufacture.

Overall supervision by the manufacturer of the quality control tasks to ensure that the quality required is obtained.

The monitoring of quality levels at any stage to provide information for management.

A management system for programming and co-ordinating in the quality maintenance improvement efforts of the various groups in a design and/or manufacturing organization so as to enable production at the most economic levels which allow for full customer satisfaction.

<table>
<tr><td align="center">AMERICAN</td><td align="center">EUROPEAN</td></tr>
</table>

Note: Broadly, quality control has to do with making quality what it should be, and quality assurance has to do with making sure quality is what it should be.

Reliability

(General). Dependability: ability of an item to perform a required function under stated conditions for a stated period of time. (As a measure.) The characteristic of an item expressed as the probability that it will perform a required function under stated conditions for a stated period of time.

The measure of the ability of a product to function successfully, when required, for the period required, in the specified environment. It is expressed as a probability.

Despite the existence of these definitions, management must realize that they are by no means universally accepted. In many cases quite different interpretations will be found to be applied to the terms, by different individuals, even in the same organization.

Unfortunately, confusion is worst in respect of 'quality assurance'. Although this is the oldest term in the business, it has recently been 'rediscovered', and its original meaning is not yet properly understood.

A few examples may be given of the lack of uniform usage. In some firms, the Chief Inspector will be found to be doing a fine job of quality control. In others, the Quality Manager is no more than a chief inspector in the old-fashioned sense. Some who have adopted Quality Control place undue emphasis upon 'control', and try to assume powers over other departments in the organization, so guaranteeing non-cooperation and ensuring the early termination of management's interest in the venture. Many people regard quality control as a specific activity closely concerned with the manufacture of the product. Some use quality assurance in the same sense. Customers' inspectorates are inclined to dignify acceptance inspection by the term, which is far too narrow an interpretation. This also hinders progress towards the understanding and implementation of quality assurance as a management system, by the supplier. When it is effective this, the original Bell concept, reduces the need for acceptance inspection by the customer to a minimum.

Management needs to be aware of this confusion of semantics and of thinking. At the same time, there is need for a clearer understanding of the roles of Inspection, Quality Control, and Quality Assurance than is conveyed by the necessarily condensed definitions listed above.

The Inspection Function

We inspect to ensure that, as far as possible, the customer receives only those goods which are satisfactory. In many cases, we inspect, too, in order to limit the amount of work which might be carried out on parts which are detectable as being already unsatisfactory, at an earlier stage in the manufacturing cycle.

In a perfect world, there would be no need for inspection. But man is fallible, and mistakes will sometimes be made. Materials and processes are variable and, in the case of sophisticated products upon whose integrity depends human life or money, it is not always easy to detect variations which might be deleterious. Hence, we inspect.

In the majority of cases, each unit of product is inspected, to a greater or lesser extent, before it is despatched on its way to the customer. The amount of inspection applied will vary from a cursory viewing for external appearance, through gauging checks on critical dimensions, to functional tests, and to the application of advanced techniques of non-destructive examination. At one extreme the inspector might be a teenage girl (and very conscientious and able she may be, when properly trained), at the other, a highly skilled and experienced technologist.

It is extremely unfortunate and symptomatic of the general lack of understanding that, for many years past, inspection has been denigrated by people who ought to know better. Many managers resent its cost. They look upon it as non-productive, if not indeed anti-productive. Production people, of all grades from operative upwards, consider the inspector to be working against their interests. The operative may have his pay packet affected, the production manager his scheduled programme, by the rejection of parts which they think are good enough. Most heinous of all are those quality control people who have promoted themselves at the expense of the inspector. They fail to realize that without inspection there can be no control of quality.

> Today's quality specialist is a technical expert concerned with all aspects of quality and their implications; inspection serves him as a measuring stick and guide post.[4]

Yet even that highly professional body the American Society for Quality Control has only recently, twenty years after its formation, established an Inspection Division.

The Purpose of Inspection The reason why we inspect is to:
  (a) Protect the company's reputation by preventing defective parts from reaching the customer.
  (b) Eliminate unproductive work by identifying and rejecting unsatisfactory components as early in the production cycle as possible.

(c) Provide information to producer, quality controller, and manager regarding the location and incidence of defective parts and the causes, and regarding the performance of individual operatives and departments.

To carry out the act of inspecting, the inspector must:

(a) Understand and interpret the specification which describes the product.
(b) Measure or judge the characteristics of the product which have been evolved.
(c) Compare the actual with the specified characteristics.
(d) Sentence the product, by accepting it if the characteristics meet or exceed the specified level of quality or rejecting it if they do not.
(e) Record the results so that production, quality, and management personnel know when, where, and if possible why trouble is occurring in the production cycle, so that remedial action can be taken.

Some Common Difficulties All too seldom is inspection carried out along the lines indicated. In most cases, there is no viable specification of characteristics and standards of quality. If a specification does exist it is usually unclear, ambiguous, or unattainable (a situation most commonly found in respect of dimensional tolerances). The inspector is therefore compelled to use his discretion and to impose his own standards, which is definitely not his job. Ask any assemblage of managers, however, and the majority of them will opine that this is what the inspector should do, thus indicating a failure to realize where responsibility lies.

The inspector cannot always know what conditions the designer has planned the product to meet, nor what standards must be achieved for the product to cope satisfactorily with its intended conditions of use. It is clearly the responsibility of the designer (assisted where necessary by quality engineering) to specify what standards must be met. Otherwise, the product is likely to be put at risk, and the designer will be deprived of valuable knowledge of the precise quality of those products which will have succeeded, or failed, to perform satisfactorily.

The absence of defined standards to be met by the production people and applied by the inspectors, coupled with the usual preoccupation with quantity of output rather than its quality, leads to cheating.

> . . . the operation of such control systems (i.e. of output) is a strong stimulus to human ingenuity (among managers as well as workers) in devising ways of 'beating the system' . . . this ingenuity is almost limitless, and the consequences are quite contrary to those devised by management.[5]

More often than not, the consequences are that quality is sacrificed in order to achieve the desired output. This merely defers the reckoning which, in the majority of cases, costs many times more than the proper action at the right time.

Who Should Inspect? So far, it has been implied that inspection is carried out by an independent and separate department. This is the most common plan, and it has been fostered for many years by the Aeronautical Inspection Directorate and similar bodies. These have stipulated that contractors shall have an inspection department which is in no way under the control of anyone concerned with output. This begs the question. It assumes, by no means always correctly, that those responsible for output are not interested in the quality of the goods they produce. It ignores the fact that in the ultimate the inspection department, and all the others, reports eventually to the Chief Executive of the Company. Is not the Chief Executive concerned with output?

One of the happiest consequences of the Q & R movement has been the growing realization that it is the production department's duty to manufacture goods to specification, and to offer for approval only those which are conscientiously believed to be good. A direct outcome of this is the further realization that the check to confirm that the production operation has been carried out properly is a direct production responsibility and charge.

Several years ago, H. W. Mander introduced into Automotive Products Ltd a comprehensive system of production-controlled inspection, coupled with a quality audit.

The activity is a large machining facility, producing important components for motor cars. The instructions are prepared by the Industrial Engineering Department, and transmitted by means of the Operation Sheet which, however, carries much more information than usual.

Each characteristic (e.g., a dimension) which is to be produced is classified as being vital, important, or as affecting appearance. There are the usual instructions to the operator, specifying the machine, the jigs, and tools which must be used.

Then follow instructions for the inspectional check which must be carried out by the production operative. These specify that after every change or regrind of a tool, or attention to the machine, the first-off part must be passed by inspection. The first part off at the start of each shift must be passed by the setter. Thereafter, the operator checks characteristics at a stipulated frequency which

185

depends upon their criticality. The gauges which must be used are also specified.

Finally, the Operation Sheet carries instructions defining the frequency and size of sample batches which must be checked by the inspector himself.

In Japan, many firms engaged in heavy, light, and precision engineering work on similar lines. In some cases, the checks are specified by the Quality Control Department, but what the department is called is of no importance. It is what is done that matters.

It is not recommended that practices such as these should be adopted without careful preparation and indoctrination of those concerned. Sudden change is seldom a good thing, and it might well take a manager a couple of years or so to set the scene for the complete adoption of quality assurance principles throughout the organization.

In the meantime, it could be a useful preparatory step to examine the existing situation as it affects the inspection department, the chief inspector himself, and his dependence upon other departments in the organization.

The Chief Inspector Owing to the false situation which generally exists, the inspector is often subject to psychological pressures of a nature not found in other activities.

When a man is transferred from a productive activity to inspection he may become imbued with a feeling of great responsibility. He gets the blame for every customer complaint about unsatisfactory product. At the same time he is blamed by production and management alike. He seldom receives help or guidance on standards of acceptable quality and he is compelled to lay down his own criteria, which will tend to vary between the too strict and the too lax.

The first step is to ensure that clear and unambiguous standards of quality are laid down. In the straightforward case of a machining activity, these will usually exist as specifications of finish and tolerances on dimensions.

The inspector should then be instructed only to accept parts which conform. Permission to accept discrepant products *must* be granted by the engineering department. Action along these lines would quickly expose the true state of affairs existing in the organization.

With the increasing complexity and high rates of production of many of today's products, with higher levels of stress, new materials and processes, the inspector is often expected to have a much higher level of technical knowledge than formerly.

186

New and sophisticated methods of measurement; a widening range of methods of non-destructive testing; functional testing of sub-assemblies; the problems of numerically-controlled machine tools; statistical sampling; new materials and processes of manufacture; the approval of the inspection systems of perhaps hundreds of supplier firms; an increasing number of official documents emanating from customer organizations: these are some of the areas of new knowledge with which the inspector is expected to be familiar.

It is no longer wise, therefore, to recruit the inspection force from among production operatives who through illness or accident require a sedentary job. It is important to realize that an inspection force possessing adequate skills, under whatever title it may operate, is essential to any successful programme of quality assurance.

Duties of the Chief Inspector It may help management to have a check list of *what* must be done. *By whom* the tasks are fulfilled will depend upon the existing organization, and wise management will avoid disturbance to an existing pattern of activity, provided that what is necessary to be done, is being done.

In general, the duties of the Chief Inspector can be outlined as:

(a) To be responsible for ensuring that all outgoing product conforms with the specifications laid down on drawings and in other relevant documents.

(b) To assist Production Planning and Production to lay down methods of in-process control of quality.

(c) To lay down methods and frequency of checks by inspection personnel.

(d) To evolve (with the assistance of Quality Engineering if such a department exists) methods of monitoring the quality of incoming materials and components, e.g., by sampling inspection.

(e) To ensure that all inspection and checking equipment, used by whatever department, is maintained in satisfactory condition.

(f) To collaborate with the Purchase Department in the assessment, education, and monitoring of outside suppliers.

(g) To provide information to the Production Manager regarding the quality performance of each production department.

(h) To assist the Production Department in dealing with problems which can be solved without diverting personnel from their essential day-to-day duties.

To Whom should Inspection Report? For the present we are considering the inspection department broadly in those firms which have not yet set up a quality control activity, as well as in those which have done so.

Some years ago, a survey was made of the quality control practices of forty-four European firms, in widely different industries. This sample was a biased one insofar as the respondents to the questionnaire which was circulated were all adherents of quality control. The survey showed that:

The Inspection reports to or is integral with

The Quality Control Department	52%
The Production Department	34%
The Technical Director	14%

Where the more technologically advanced British firms are concerned, present indications are that inspection *per se* is increasingly reporting to the Production Department or to the Technical (i.e., Engineering) Department.

The Operative/Inspector Ratio It is important that management should realize that the inspection function may well absorb a significant proportion of the total personnel involved in output. On average, the inspection force is about 15 per cent of the direct labour force.

The ratio of direct operatives to inspectors is a popular measure with many managers, who use it to encourage their chief inspectors to reduce the size of their departments. It is a ratio, however, which should be used with great care.

The ratio may safely be used to compare the efficiency of departments carrying out similar duties, in a similar way, in a single organization. In general, however, the figure varies widely, under a number of different influences. These include the extent to which the work is automated or tooled-up; the rate and scale of output; the penalties of inspection failure; the relative skill of operatives and inspectors; their gender; the complexity of the product; the extent to which inspectors must make checks of material cleanliness, as by non-destructive testing; the extent to which the production operative has been trained and motivated to accept the responsibility for producing good work and for monitoring his own performance.

In extreme cases the ratio has been found to be 1·2:1 (for a high class ball bearing) and 100:1 (for ice-cream and frozen foods).[6] Clearly, if a process for the manufacture of a product having critical characteristics could be completely automated, the ratio might approach the inverse of infinity. Indeed, at the time of writing a current advertisement by British Gypsum Ltd reads that the company has 'more inspectors than production people' on their manufacture of plaster and plasterboard, representing a ratio of less than unity.

188

The wide divergencies in different industries, indeed within one type of industry, are shown in Fig. 14.1.

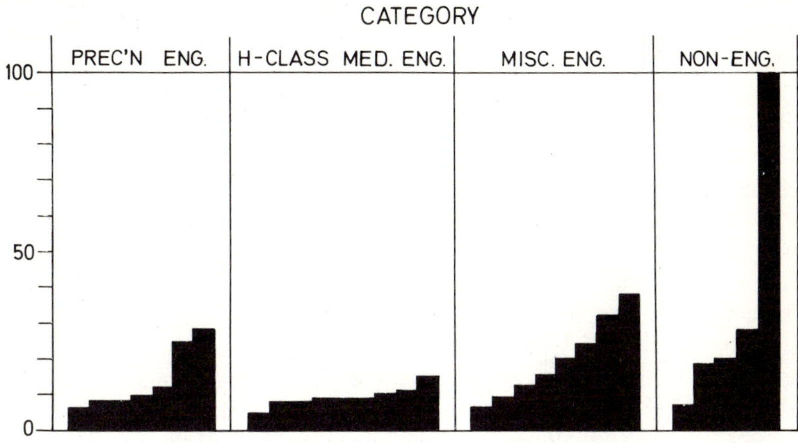

Fig. 14.1. Operator/Inspector Ratio.

Current figures for Japanese engineering firms where the operatives check their own work suggests that the following ratios, between operatives and quality control personnel, are not unusual:

automobile engineers   20:1
aircraft engineers   between   6:1 and 12:1
cameras   12:1

The Function of Quality Control

We shall avoid the confusion which still surrounds quality control if we remind ourselves that the term is a condensed version of 'economic control of product quality' or of 'in-process control of product quality'.

These terms reveal the essential difference between inspection and quality control. Inspection is by its very nature a post-mortem activity. A product is, at the point of inspection, either satisfactory or not. The operations which have resulted in the present state of the product are over and done with.

A prime purpose of quality control is to ensure that operations of manufacture are in control, so that satisfactory product is the outcome. It will be evident from this statement that this does not necessarily require a new professional approach. Much will depend upon the circumstances prevailing in each individual firm.

It has been emphasized repeatedly that there can be no standard or universal pattern of the implementation of the principles of quality

189

control and of quality assurance. It is *what* is done, not *how* or by *whom* that is of importance.

It is unfortunate that some Government departments stipulate that a contractor, to become 'approved', shall set up a quality control department. As we have seen, this is in marked contrast to the practice of the US Department of Defense and of some of the largest companies in Britain. They lay down what is required of a supplier. How he does it is his business.

Management will, therefore, be well-advised to take careful stock of its own situation. Often it will be found that sound practices already exist. Indeed, how else would the firm have remained viable? The necessary activities may be carried out under some title not recognized in the current jargon, or, more probably, they may not have been established on a formal basis. In such a case the soundness of the organization may not be recognizable to a modern cultist. But to inject into such a company a quality control department, with the responsibilities described in some of the official 'Requirements' could be to sow the seeds of inter-departmental conflict and to frustrate the aspirations of those 'Requirements'.

It is strongly recommended that, before any change is made, a list should be prepared showing the present functions and duties of existing departments. A check should be made that the personnel concerned agree that these responsibilities are theirs, and that they carry out their functions to the satisfaction of the other departments.

Often will it be found that there are areas where there is incomplete understanding or incomplete information or poor communications. In the smaller firm it should be relatively easy to make good these faults, after which the efficiency of the organization should be improved greatly. In the larger organization, to set up a quality control department may be the right decision. The smaller firm should realize, however, that the unconsidered copying of larger firms is not necessarily the right action for them to take.

Some years ago the present writer prepared the chart of activities reproduced in Fig. 14.2. This was widely distributed in the hope that constructive criticism would be received, so that a comprehensive and generally applicable guide could be prepared. Unfortunately, it was accepted with uncritical enthusiasm, and it has been left to its author to criticize his own work.

In retrospect, the superscription is felt still to be appropriate. With increased experience, however, one regrets (a) the failure to emphasize the importance of producing goods to meet the specification, and the responsibilities of those concerned with planning, procurement and

190

Requisite Quality is defined as that which is necessary to ensure customer satisfaction, by achieving a given standard of reliability, or fitness for purpose, in the product.

The exercise of full control of product quality will require the application of common sense, engineering, inspection, proving (e.g., by testing) and statistical methods of analysing manufacturing performance, each in greater or lesser degree according to the nature of the product, its value, its purpose, the rate and scale of its production, the personality of the organization, and the capabilities of the personnel concerned. The chart indicates the basic acts which must be carried out for full control to exist, but these acts will vary considerably in individual importance according to the needs of the enterprise under review.

There is no simple universal solution, nor any single ideal form of organization. The precise title under which each of the basic acts is operated is of no significance, and it may well vary in different organizations.

The chart should be used as a check list to ensure that no essential step has been overlooked, and as an aid to the diagnosis of the needs of the organization. Items marked * are definitely the responsibility of Inspection and represent the great bulk of the total personnel required. See text of Paper.

CONTROL OF PRODUCT QUALITY CONSISTS OF

1.	2.	3.	4.
ENSURING THAT ADEQUATE SPECIFICATIONS EXIST,	CHECKING THAT THE SPECIFIED STANDARDS ARE MET,	INVESTIGATING AND ELIMINATING CAUSES OF NON-CONFORMANCE,	AS CHEAPLY AS POSSIBLE.

1.

The overall Quality concept requires that the Requisite Quality of Product shall be carefully determined, and fully and clearly defined.

Too high a standard leads to loss of profit, too low a standard to loss of reputation and business.

This function requires a study (a) of what is needed to enable the product to fulfil its purpose, and (b) of the variability likely to arise in quantity production. Reliability and customer satisfaction will ensue only when the range of quality is such that the lowest level will always exceed that which is required of the product.

1.1 Study the needs of the product in the light of all available knowledge. Analyse records of prototype trials and of service complaints of similar products, to determine necessary level of quality.

1.2 Determine variability likely to arise in production, by observation, analysis of plant capability, rig testing, life- or fatigue-testing, as appropriate.

1.3 Ensure that Acceptable Quality Level is specified clearly and unequivocally, based upon the necessary level of quality determined from 1.1.

1.4 Check Engineering Drawings, Specifications of Materials and Processes, and Quality Specifications, to ensure that together they convey to the manufacturer the A.Q.L. of 1.3.

1.5 Develop methods of inspection, non-destructive testing, and proof testing as necessary to enable the A.Q.L. to be achieved.

2.

This is the responsibility of the Inspection Department as it is commonly understood, excepting that the Inspection Audit function should ideally be carried out by an independent body.

*2.1 Prepare Inspection Plan, stipulating stages in production process at which inspections shall be carried out, and defining method and equipment.

*2.2 Check production equipment, e.g., machine tools, jigs, and fixtures. Are these adequate to enable the required standards to be met?

*2.3 Control operations and processes to the standards laid down.

*2.4 Check the product against specification.

*2.5 Accept or reject the product.

*2.6 Record reasons for rejection.

*2.7 Overcheck systems and organization of Vendors and Sub-contractors.

*2.8 Overcheck products of Vendors and Sub-contractors.

*2.9 Check ability of operators carrying out critical work, e.g., welders.

2.10 Check efficiency of inspectors.

2.11 Conduct Inspection Audit or Quality Assurance checks on finished products.

3.

Inspection and Production personnel will naturally take all possible action to eliminate causes of scrap, where these are appreciated and can be diagnosed quickly. Many problems will require a fuller investigation than can be made by those concerned with the day-to-day running of the Shop, however, and it will usually be found more expedient and effective to set up a small Quality Engineering activity with time to study these problems thoroughly.

3.1 Set up system for the rapid and accurate recording of rejections and rework.

3.2 Investigate causes of rejection and rework.

3.3 Feed back information to Production, Purchase, or Material Supplier for remedial action as indicated.

3.4 Feed back information to Design for drawing change where necessary.

3.5 Develop new methods of measurement and Non-Destructive Testing as required for earlier identification of defective work.

4.

Although action along the lines already indicated will result in considerable economies, Minimum Cost for Requisite Quality can only be achieved by regarding the overall cost of Quality Assurance, i.e., Inspection, Testing, Scrap and Rework as involving a definite proportion of the total manufacturing cost which can and should be reduced.

Every facet of the activity should, therefore, be kept under constant surveillance and treated as a productive activity, the efficiency of which can be increased by the application of modern management techniques.

4.1 Determine total cost of quality assurance, i.e. Inspection and Quality Control labour, equipment, scrap and rework, testing.

4.2 Determine stages at which scrap arises. Redeploy inspection force to increase its effectiveness at critical points in manufacturing programme.

4.3 Do not increase inspection effort beyond the point of economical return.

4.4 Conduct tests on rejected parts to determine acceptability on a concessionary basis.

4.5 Develop processes to improve capability at no increase in cost.

4.6 Develop more efficient and automatic methods of manufacture and inspection to reduce the risk of human error.

4.7 Reduce amount of inspection by use of Control Charts and Sampling Schemes, where possible.

4.8 Apply Work Study to Inspection operations.

Fig. 14.2. A Guide to the Control of Product Quality.

production for doing this, (b) the over-emphasis upon the quality control function.

Management which has taken the trouble to elucidate the situation will be in no doubt as to where responsibility lies. Management alone can take the necessary action to remove causes of friction and inefficiency. It will be clear too that until this has been done, the strengthening of the inspection department, or the setting up of a quality department to ensure compliance with a large customer's requirements, will avail little. Indeed, it may well make matters worse since it can provide an alibi to those departments which have been most at fault and which it must be management's concern to involve in the total activity.

Before defining the scope and responsibilities of its new quality department, senior management should consider the foregoing, and in addition pay heed to the advice of John E. Condon:

> One of the major factors often overlooked by industrial management in dealing with the quality function is the inadvertent or unconscious failure to recognize that the quality assurance function does not have a creative role relative to the hardware.
>
> The creative role is with the designer and the manufacturer. Quality's role is to see that, in fact, the part is built to the established requirements. If the requirements are wrong, responsibility goes back to the designer. If the hardware is not manufactured in accordance with the right requirement, responsibility goes back to manufacturing.[7]

Management could not wish for clearer advice. In case they should feel, however, that this is only for the space industries, not applicable to their more earthly activities, let them beware. The advice is directly applicable to every manufacturing activity. Topically and particularly, it is likely to have an early impact upon thousands of manufacturing firms, as a direct consequence of the US National Traffic and Motor Car Safety Act, 1966. Only slowly is it being realized that this Act is going to affect a far wider circle of firms than just the motor car manufacturers themselves. Each main firm may depend upon from 700 to 1000 sub-contractors and vendors. Large suppliers of specialist equipment, such as the Lucas Organization and Automotive Products Ltd, may in their turn depend upon a like number of suppliers. Each of these subsidiary firms will depend upon at least 100 more suppliers. The stringent requirements of the Act regarding the integrity of specified motor car components having an influence upon passenger and third-party safety will, sooner or later, affect the design and manufacturing

192

facilities and practices of all these companies. They will penetrate into the farthest corners of industry.

Lest it be thought that the foregoing warnings and provisos should leave a quality department with nothing to do, it will be interesting to consider an actual example.

A Medium-Sized Firm's Adoption of Quality Control The chief executive of a firm of 3000 people engaged in mechanical engineering, with a range of products varying between the rough and heavy and light precision, had had an unfortunate first experience of quality control. Fortunately, he remained convinced of its value, and sought advice.

The pros and cons of setting up the Quality Engineering Department under either the chief engineer or the general manager were carefully considered. It was decided that the latter should take the responsibility, because he had the wider experience, and could take the broader viewpoint. The Quality Engineering Department was established under the control of a quality manager, an experienced graduate engineer, who had reporting to him a chief quality engineer, a chief inspector, and a laboratory manager. His terms of reference are set out below.

QUALITY MANAGER

Purpose: To form a quality control organization and coordinate its work.

Duties and Responsibilities:

1. Management of the quality control organization.
2. Maintenance, as far as possible, of product quality levels consistent with divisional policy on quality and within the framework of approved expenditure.
3. Provision of assistance to all other departments to enable them to carry out their responsibilities towards quality.
4. Recommendation where appropriate for the introduction of new, and the alteration of existing quality standards.
5. Advice on the maintenance of product quality at minimum overall cost.
6. Maintenance of direct communication links with those responsible for quality at company headquarters and other divisions of the company.

CHIEF QUALITY ENGINEER

Purpose: To form and manage a quality engineering department.

Duties and Responsibilities:

1. The organization and management of the quality engineering department and the fulfilment of its function.

2. Agreeing to, and working within, an annual budget for expenditure.
3. Maintenance of a close working relationship with the divisional sales, service, and engineering departments in order to assist in rapid feedback from the customer and effective remedial action on the company's products.
4. Maintenance of close and effective communications with customer's resident personnel involved with the quality of components manufactured by the company.
5. Ensuring close liaison with all departments to maintain effective cooperation, interchange of information, and advice for the control of product quality.
6. Maintaining lines of communication and exchange of information with associated organizations in other divisions of the company.
7. Keeping abreast of new developments in the field of quality and making recommendations to incorporate these, or other changes where appropriate and worthwhile.
8. Assisting in the preparation and presentation of information for short- and long-term proposals and projects.
9. Arranging, in conjunction with the personnel and training department, suitable selection, instruction, and training procedures to ensure a high level of efficiency and effectiveness of facilities and personnel, and adequate promotive resources.
10. Maintaining, at all times, a close working relationship with the chief inspector and the laboratory manager and ensuring that personnel at all levels in their respective departments, similarly cooperate.

The Quality Engineer and/or Manager The selection of the right man to develop and run the quality activity is vital to its continuing success. The easy solution, which, unfortunately, is encouraged by some of the guides and requirements published by Government departments, is to change the title of the chief inspector, appointing him to fill the quality post.

It is in no way criticizing or denigrating the office of the chief inspector to say that only in exceptional cases will this be the right action. No matter how sound a man he may be, the chief inspector will start with considerable disadvantages. In many cases, he will have been looked upon by production as the enemy of output. By engineering, he will be thought of only as the man who is responsible for most of the customer complaints. Added to the difficulties of establishing a new activity, therefore, would be the need to live down a background with which he had been identified for many years. In addition, he would almost certainly be oriented towards manufacturing. He would experience difficulty in communicating with design engineers and management.

Of course, there will be exceptions. Indeed, there are notable examples of chief inspectors who have established excellent quality control departments. Management which is sincerely anxious to achieve the best performance from the whole organization—and this means complete coordination and integration—will, however, be well-advised to resist the temptation to appoint the chief inspector merely because he will have made a study of and pressed for the adoption of quality control.

The long-term good of the company depends upon bringing more closely together the engineering and manufacturing departments. The quality executive can help considerably towards this end. The contribution which he can make should therefore be an important consideration in the selection of the man for the job. Obviously, he should be experienced in the traditions and methods of the company. He should preferably be a trained engineer. Whether or not he should have a production or an engineering background is probably of secondary importance, provided that he possesses the attributes of persistence and tact, and is capable of preserving a broad outlook. Paterson, whose experience exceeds that of most of us, makes an interesting comment:

> Despite its origin in an R & D organization (Bell Laboratories) and its growth and continued close association with the broad activities of its birthplace, quality control is often defined and treated as part of manufacture (probably because the bulk of the literature and much of the practice is limited to manufacturing). Such a circumscription is an unwarranted limitation of the intent of Shewhart, who conceived the term; it also tends to stifle what should be Q.C.'s earliest and one of its most important functions— the reconciliation of the interface between design and manufacture.[4]

Although it is not mentioned specifically, it should be remembered that the Bell approach was always the total one, encompassing reliability as well as quality. In firms where reliability is important, it may prove to be best in the long run to appoint a man as quality engineer who has experience of, and who is acceptable to, the engineering department.

There is no need to have misgivings about the effect upon the chief inspector. As has been shown, his knowledge and experience will be essential to the new department, and as it establishes itself, his importance will be enhanced.

Establishing the Philosophy and the Practice Having gone so far as to set up a quality department, with clear terms of reference, management might be inclined to sit back, thinking that it had completed its share of the task. More is required, and the successful projects are identifiable as those where senior management has gone on to establish a permanent system of reporting and accountability.

Allowing time for the new department to establish itself, the next step should be to ask that it should make the attempt to define the contributions and responsibilities of all the departments of the organization. This information should then be taken as the basis for a Quality Manual. This should make clear to all the departments of the organization the responsibilities which they bear towards the quality and reliability of the end product.

An excellent example of a Company Quality Manual is that prepared by Mander for the Automotive Products Group of companies. It differs from those manuals which are produced merely to satisfy the requirements of a major customer in that it involves every department of the organization in the overall objective.

The Manual defines the organization and the responsibilities of Central Quality Control, which consists of a small, qualified and experienced staff, comprising the Group Quality Manager, the Quality Methods Engineer, the Q.C. Instructor, a statistical quality engineer and his assistant, and supplier quality assurance officers. As a matter of routine, a meeting is held with design engineering whenever a new design approaches the detail design stage, to consider the quality and reliability hazards which might be present in the design. The production engineering department (industrial engineering) collaborates with quality control to produce Operation Sheets which specify checks by the setter and the operator, and a quality audit. It is laid down that the production department is responsible for producing work to specification. Patrol inspectors record operator performance on simple control charts. The assurance of quality of external supplies is operated by a committee consisting of buyers, chief inspectors, laboratory specialists, and the supplier quality assurance officer, under the chairmanship of Central Quality Control. Progress is monitored at a monthly Divisional Quality Meeting, which is attended by the production manager, the chief engineer, the chief industrial engineer, the divisional quality engineer, the method study engineer, the planning engineer, and the production service engineer. The chief inspector takes the chair. At this meeting there is presented a monthly report showing divisional quality costs.

An important feature of this company's activities has always been the training of people in all activities within the organization.

It will be appreciated that the nature of this company's work, as a major supplier of proprietary items to the motor car industry, tends to

196

be product-oriented. The point to note is that the Quality Control Manual involves each activity in the organization.

The way in which this Manual was planned has been described in a useful article by Holmes.[8] He has combined a general review of manuals with a good deal of practical advice.

A Spanish company, Metalurgica de Santa Ana SA, has had a manual of quality policy and practice for several years. By kind permission of Dr Jesus Garcia del Valle, the Director of Engineering and Quality Control, who is its author, it is possible to reproduce the relevant portions of the latest version in Appendix 1. It will serve as a guide to others intending to produce a manual for themselves.

The Place of the Quality Department in the Organization A question which is commonly asked by company executives is 'Where should I place responsibility for quality?' For many years, Government Inspectorates have stated that the chief inspector (who, they say, may sometimes be known as the quality manager!) should not be responsible to any executive who is also responsible for output.

This is rather begging the question. It assumes that no executive is capable of or willing to produce satisfactory product. It shows a lack of appreciation of current trends in management thinking. It does not see that every department is responsible ultimately to the chief executive, who must, himself, carry the responsibility for the total performance of the organization of which he is the head.

Experience shows that there is some justification for this assumption, for there are some firms in which quantity is still allowed to transcend quality. But the role of a Government department should surely be to lead its suppliers towards clearer appreciation of the fundamentals of their joint business enterprise. Only thus can overall efficiency, better value for the taxpayer's money, be achieved. To perpetuate the policing role must inevitably result in increased costs which the more advanced sectors of industry have already shown are avoidable.

Many of the more forward-looking firms today, especially those which have understood and accepted the principles of Q & R, realize that the responsibility for producing goods to specification rests primarily with the manufacturing division. In such companies, as we have seen when discussing the practice of Automotive Products Ltd, and some Japanese firms, the act of inspection is considered to be a necessary part of the production process. A separate inspection department is either non-existent already, or rapidly diminishing.

In those companies where this trend has been recognized and properly understood, there is a tendency to put the inspection department under the production executive. There is nothing wrong with this, indeed it is to be welcomed, when it is clearly recognized that it is the job of manufacturing to produce components and products which comply with the specification.

As regards the Q.C. department itself, what it does, and what authority it exercises, are of greater importance than to whom it reports. Its function, its role and its position in the organization will depend upon the extent to which the manufacturing division has accepted its responsibilities. For this reason and others, the chain of command will vary from firm to firm. The senior executive most suitable to take the quality activity under his wing will obviously be the one who has the best understanding of the role and the function of the department, and who is prepared to give it his full support. Where this executive stands in the organization will depend upon the size of the firm, upon the status of the heads of other departments, and upon the status relative to these, which it has been decided should be accorded to the quality activity.

Hence, there can be no uniform pattern, and only the chief executive and his executive directors can decide, from their knowledge of the organization and its needs, and of the personalities, the tensions and the pressures involved, where the Q.C. department should be located so as to have the best chance of survival and success.

Despite what has been said, an examination of existing organizational patterns may be found useful to managements anxious to establish their Q.C. department on the best possible basis.

The analysis which was made of the practices of 48 firms in 8 different European countries[6] showed that the Heads of Quality Departments reported as follows:

> To: The Board of Directors, or General Management   46%
>   Production Management   34%
>   Technical Directors or Management   8%
>   The Directors of Quality reporting to Board   2%

Were the survey to be repeated today it is felt that there would be revealed an increasing tendency to make the quality activity, under whatever title it operates, equivalent in status to the other major activities in the organization. This is shown in the typical organization charts in Figs. 14.3 and 14.4.

What's in a Name? It may have been noticed that no firm line has been taken to adopt one title or another for the head of the quality control activity. At the time of writing, 'quality manager' appears to be in the ascendancy. If the organization happens to be technically-oriented, and

198

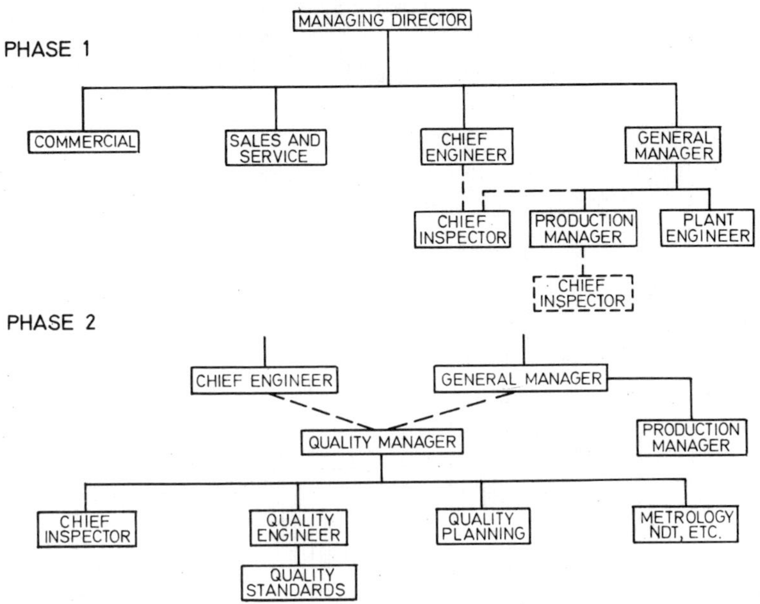

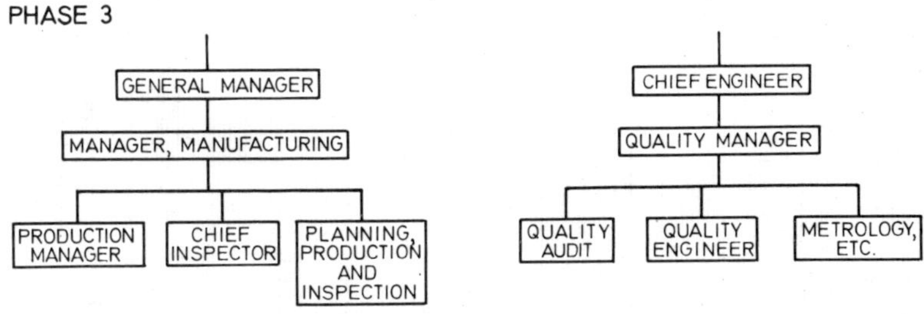

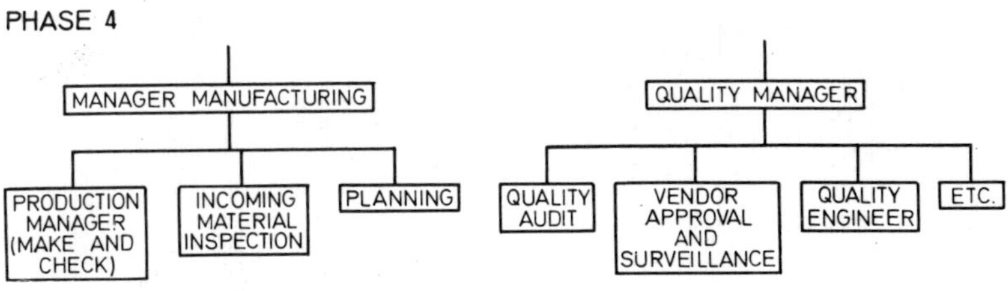

Fig. 14.3. Typical Organization: Single Plant, Single Product.

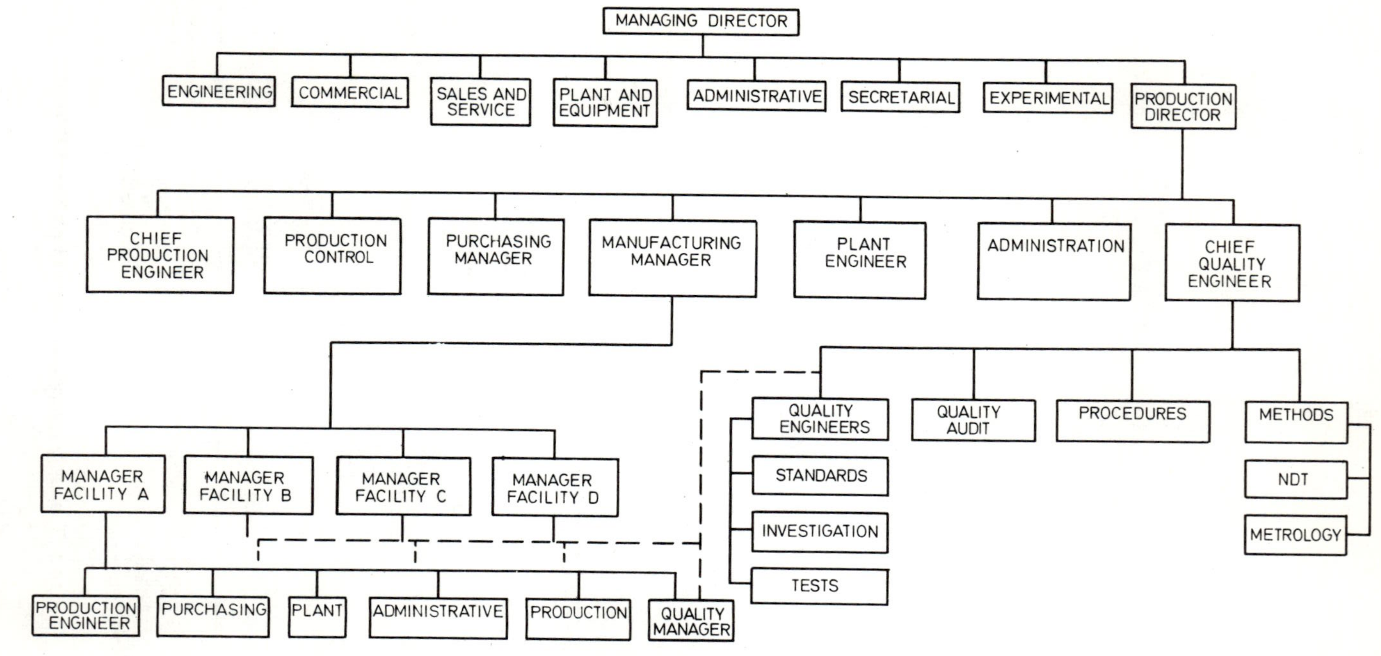

Fig. 14.4. Developed Organization for Multi-Plant Operation.
(Dotted lines indicate 'staff' responsibilities)

if the head of the department has been recruited from the engineering department, he may prefer to be called a quality engineer. In some firms, the quality manager has quality engineers reporting to him. In others, the reverse is the case. It does not really matter. What is important is that the individuals should feel happy, and that they should know precisely what they are expected to do.

Quality Assurance

When management can be satisfied that its inspection and quality control activities have been established on a sound basis, it will be ready to take the ultimate step. This is to set up a system of procedures, instructions, and records which will ensure that each department in the organization carries out its duties towards the achievement of good end-product, and can be seen to have done so.

This requires action by top management, since it involves leadership and the involvement of everyone in the whole complex enterprise. It is quality assurance as described in the new ASQC definition quoted earlier. It is the total activity which it is the aim of the British Navy Department, the US Department of Defense, NASA, and NATO to achieve throughout the industries supporting them.

As we have seen, quality assurance in this sense is essential to the achievement of reliable product, although it is not so vital where the end-product is non-functional and where its quality can be confirmed by an inspection.

It was the realization that the reliability of a product depends upon the extent to which very many processes and operations are carried out correctly which led to the system established 50 years ago by the British Air Ministry. Riordan has described the basic system, tracing the developments of this 'new' approach in the United States from its beginnings in 1954.[9]

1. The quality of manufactured products is dependent primarily upon the effectiveness of the manufacturer's control over fabrication, inspection, and testing operations.
2. In consequence, manufacturers are responsible for instituting such controls of operation, processes, and checks as are necessary to ensure that the products conform to the customer's requirements. The manufacturers are also obliged to provide objective, verifiable evidence that they have carried out all the necessary activities.
3. This means that a producer is expected to sell to a customer not only products and services but, in addition, proof that the product has been properly made and tested.

201

4. It is recognized that the customer cannot usually protect himself by inspecting the finished product, save in those cases where 'quality' is the only desired characteristic, as of writing paper, a piece of furniture, a bottle of wine. Where life, durability, or reliability are the characteristics desired, only time and experience can establish their goodness. A fair measure of assurance can be gained, however, by ensuring that everything necessary has been done to achieve the required integrity of each component of the finished product.

5. For his part, the customer must define clearly his own requirements, which are categorized as:

    (a) Product specifications—defining the characteristics of the product,
    (b) Systems specifications—details of the environment with which the product is expected to cope, and the tests required to establish the product capability.

6. It is incumbent upon the customer to exercise surveillance over the manufacturer's controls, inspection and testing to assure himself that the supplier has met his part of the contract. 'Minimum surveillance is adequate when a producer is demonstrably reliable.'

When confronted with such unassailable logic, expressed so clearly, it is difficult to see how anyone can presume to attempt to justify large inspectorates and least of all the intrusion of third-party surveillance between supplier and customer.

For the concepts to be effective it is of course necessary that the requirements and responsibilities should be defined in the business agreement, the contract, between supplier and customer. This needs collaboration between all those concerned, on both sides—management, purchasing officer, and quality manager of both parties, and there must of course be clear specifications. It is only the absence of some or all of these which permits the third-party intruders to gain a foothold.

These matters have been dealt with at some length because they are fundamental to the assurance of product quality in whatever field. With only slight modification of wording the specifications could apply to the affairs of civil engineering contractors as much as to large-scale retailers like Marks and Spencer Ltd and Sears Roebuck and Co.; in fact, to any customer-supplier relationship. It is interesting in this context to note that Sears Roebuck were quite early informing their suppliers of their own definition of quality assurance, which is thought to be better than later definitions evolved by organizations having their roots in inspection and quality control. It reads:

> Quality Assurance: the comprehensive direction of all phases of product quality, from inception until the product has served its useful life. 'Cradle to grave' quality control.

Can senior management doubt any longer their responsibility towards the achievement of quality and reliability of product?

References

1. Riordan, John J. 'Cults, "Ilities", and systems management.' Defense Management Journal, Directorate of Cost Reduction and Management Improvement Policy, Department of Defense, Washington D.C., U.S.A., Spring 1969
2. 'Glossary of general terms used in quality control.' Proposed 1969 Revision, ASQC Standard A3. Quality Progress, July 1969.
3. 'Glossary of terms used in quality control.' 2nd Edn. European Organization for Quality Control, 1969.
4. Paterson, E. G. D. 'Assuring quality.' International Science and Technology, September 1963, New York.
5. McGregor, D. 'The changing nature of the managerial task.' British Institute of Management, October 1961, London.
6. Nixon, F. 'Quality reporting, quality auditing and quality achievements in European countries.' Proc. Vol. II, VIIIth Congres de l'EOQC, 1964, Baden Baden, Germany.
7. Condon, John E. NASA Quality and Reliability Assurance Office. Interviewed, *Quality Progress*, July 1968.
8. Holmes, J. 'Quality manuals.' *Quality Progress*, June 1969.
9. Riordan, John J. 'Contractual relationship between Government and contractors in quality assurance.' The Conference on Statistical Quality Control Methodology in Highway and Airfield Construction. University of Virginia, Charlottesville, Va. May 1966.

PART 5

Ensuring Customer Satisfaction

15

Product Support

No matter what pains may have been taken during the design and development phases to ensure that the product will be capable of meeting all the demands likely to be met in use, it is more usually than not the case that 'teething' troubles will be experienced when a new product enters service.

The conscientious manufacturer will have spent large sums of money in the effort to make the production version of his product as reliable as possible. In the case of aircraft engines, aircraft, and electronic equipment, the launching costs—the cost of research, design, development, and tooling[1]—may amount to between 100 and 300 times the cost of unit product.[2] Regrettably, there are many other cases where factors militate against the expenditure of the time and money necessary to establish a viable product. A small company may not possess the financial resources. There may be a seller's market, which condition is apt to engender a measure of cynical opportunism, especially in the minds of those concerned with expenditure and income. There may be insufficient time because a competitor is winning the market.

In every case, however, it is expecting too much to hope that every condition of use will have been anticipated and dealt with. It can almost be guaranteed that some user, more intelligent or more moronic than average, will find some unexpected way of misusing or abusing the product. Even in the case of aircraft engines, where millions of pounds will have been spent on the effort to avoid trouble, the hoped-for-reliability is not always realized when a new type of engine first enters service. With other types of product, where insufficient time has been allowed for design, and especially for development testing, trouble in service is fairly certain to arise.

We have seen how great can be the costs to the user of failure of plant or equipment to operate satisfactorily, on time (see page 22). In the case of large-scale capital equipment, the losses are often suffered by one single user. With domestic equipment the individual loss or inconvenience may be much smaller, even though the aggregate losses may be as great.

Whether the customer is a large corporation, a public utility, or a single individual, the ethics and the principles are the same. The customer is entitled to receive satisfaction. In the event that the product fails to give the expected satisfaction, it is morally incumbent upon the supplier to diminish the degree of dissatisfaction (i.e. of inconvenience), experienced by the customer. Action should, therefore, be taken to keep the equipment in operation, by replacement of defective components where necessary, by 'quick-fix' remedies when possible.

It is also good sound sense on the part of the producer to do all he can to deliver 'customer satisfaction'. For if he does not, although he may enjoy a share of the market for a time, the potential customer will sooner or later learn to discriminate, and in the long run the firms producing best value will be the successful survivors. This is well understood by Japanese Airlines, who state in their advertisement '"O-Kyaku" is the Japanese word for both customer and guest'.

Individual motorists who suffer from quite unnecessary unreliability, and who receive little sympathy or help from dealers and producing firms alike, may remain unconvinced by these arguments. There are strong indications, however, that increasing competition from overseas will offer the customer greater choice. There is, too, an important move towards better education of the customer, by consumer associations. The report issued by the Swedish Motor Vehicle Inspection Company itemizes the faults found during 2 million compulsory vehicle inspections carried out on 111 different 1965 models of cars and lorries. The faults are classified in the Pareto diagram in Fig. 15.1.

In the counter direction, there is a tendency on the part of manufacturers of electrical domestic equipment to 'rationalize' their designs into a smaller number of integral components. This has the effect of greatly increasing the cost to the customer of replacement parts. Whether or not it increases the total cost of ownership throughout the useful life of the product does not appear to have been taken into account.

The opinions of the users, however, can hardly be in doubt. Those who own motor cars made by a firm whose policy is to offer low priced replacement parts, with standard fitting charges, will be particularly liable to make adverse comparisons of other makes.

Such attempts as are made to placate dissatisfied customers usually

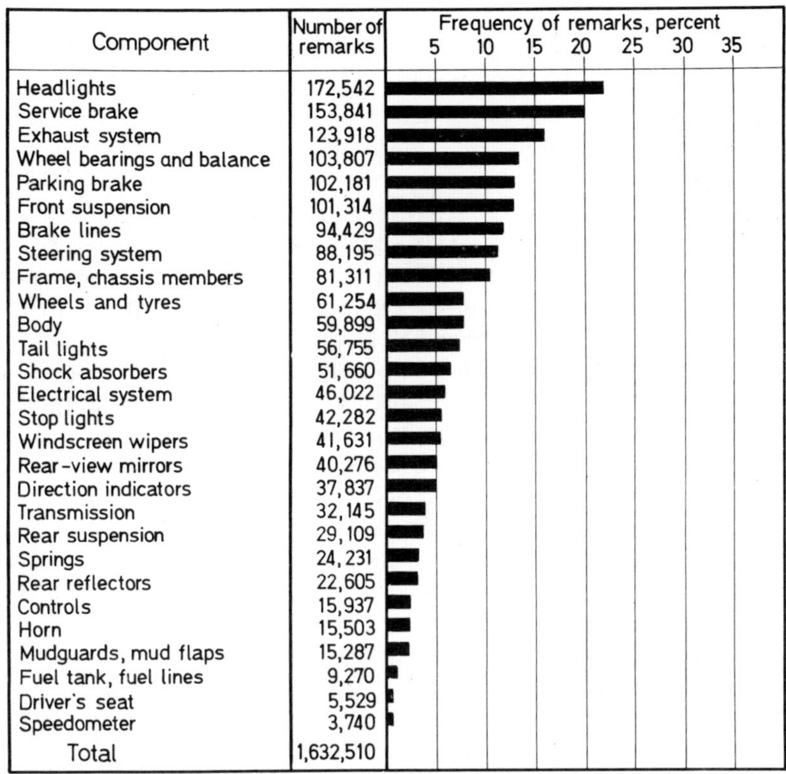

Component	Number of remarks	Frequency of remarks, percent
Headlights	172,542	
Service brake	153,841	
Exhaust system	123,918	
Wheel bearings and balance	103,807	
Parking brake	102,181	
Front suspension	101,314	
Brake lines	94,429	
Steering system	88,195	
Frame, chassis members	81,311	
Wheels and tyres	61,254	
Body	59,899	
Tail lights	56,755	
Shock absorbers	51,660	
Electrical system	46,022	
Stop lights	42,282	
Windscreen wipers	41,631	
Rear-view mirrors	40,276	
Direction indicators	37,837	
Transmission	32,145	
Rear suspension	29,109	
Springs	24,231	
Rear reflectors	22,605	
Controls	15,937	
Horn	15,503	
Mudguards, mud flaps	15,287	
Fuel tank, fuel lines	9,270	
Driver's seat	5,529	
Speedometer	3,740	
Total	1,632,510	

Fig. 15.1. Ranking of Defects Found in Compulsory Checks on Motor Cars in Sweden.

come under the heading of after-sales service. It is not realized as widely as it should be that this department has, or should have, two separate activities: product support, which covers the acts necessary to minimize the impact upon the customer of unsatisfactory product, and product improvement. This last is the other important role of the effective service department: to collect, analyse, and present information regarding the behaviour of the products in service, and to press for design improvements.

Product Support

The purpose of the product support activity of the service department is to enable the customer to get maximum value from the product. The

209

size and the technical strength of the department will depend upon the technical complexity of the product, its reliability, the consequences of failure, and the extent to which the users are dispersed over the world.

With aircraft and aircraft engines, there is the paramount need for safety. In civil aircraft operation, large financial losses can be suffered through non-availability of aircraft because of unreliability. In the military field, considerations of defence require maximum availability and readiness of the equipment. In both cases, a supplier's customers may be spread over the globe. A large aircraft engine company may in consequence have a service department amounting to 2 per cent or more of its total strength (i.e. 1000 or more people), with skilled engineers located permanently in countries using the firm's products. The service department of a motor car firm will be very much smaller in proportion, and a high degree of dependence may be placed upon accredited dealers, to act as the firm's agents. Much domestic electrical equipment is dealt with by local electricity authorities, or by retailers. The need to provide some measure of service is recognized.

In many cases the liability of the supplier is usually limited to a period of 12 months, and often even then only to the purchaser who has signed and returned his warranty card. In many instances, it is necessary to return the failed equipment to the manufacturer, at the user's expense. It is not surprising that many domestic users find this to be more trouble than paying a local dealer to carry out a repair. This latter course often has the additional merit of speed, an important matter when a man wishes to see an important football match on his television set, for example.

While it is true that some reputable dealers would be prepared to lend a television set to the unhappy customer, until the repair had been completed, such an arrangement is usually entirely local. It does not involve the manufacturer directly. This basic lack of service support is an important factor leading to the trend towards the hiring rather than the purchase of television sets. All that the domestic user can hope for is that the example set by the Chrysler Motor Company, in offering a warranty of 5 years or 50,000 miles on their motor cars, will be followed by other firms and other industries.

Where a user derives direct financial benefit from equipment, he is inclined to take a tougher line with his supplier. So we see extremely effective service departments operated by manufacturers of food- and beverage-dispensing machines, and one-arm bandits. Equipment out of action results directly in loss of profits, and this is clearly seen and felt.

On the other hand, there remains considerable scope for a better realization of their responsibilities by manufacturers of machine tools and processing plant. Many of them seem to be oblivious of the difficulties

210

of their customers, whose whole complex programme of output may be seriously upset by the non-availability of a relatively minor component, a new design of pump, say, upon which an expensive machine-tool, or even a whole chemical plant, may depend.

The prime responsibility of the service engineering department can be stated as being to help the user to get the best value from his equipment by:

(a) Technical documentation.
(b) Education of the user.
(c) Skilled and prompt attention to all customer complaints.
(d) The planning and provisioning of adequate stocks of spare parts and assemblies, to ensure rapid replacement of defectives.
(e) 'Quick-fix' remedies to keep the equipment operable until a proper remedy becomes available.
(f) Preventive or corrective maintenance, to keep the whole equipment operable, by cleaning, lubrication, repair, or replacement of components having a shorter individual life than that of the total product.

Although much more needs to be done than is still usual in most cases, the cost need not be great. It will in general be more than justified by the greatly increased customer goodwill, which will improve competitiveness while at the same time providing the manufacturer with invaluable knowledge regarding inherent defects in the product. This last point is dealt with later. Meanwhile, the activities just enumerated need to be described in more detail.

Technical Documentation
Every designer plans his product to meet estimated conditions of use and environment. For the product to have the best chance of behaving satisfactorily it is necessary that the user should be informed of what treatment is expected of him. Machinery needs to be kept clean and properly lubricated. It needs to be operated within its designed strength, speed, and capability.

The intelligent user who wishes to get the best life and performance from his purchase will be anxious to know how to treat it. The wise manufacturer will try to describe this, simply, clearly, and in understandable terms. Unfortunately, the standard of writing in operation and maintenance manuals is not always high. Local or company jargon is sometimes used. Where products are sent overseas there have been complaints that the manuals have been available only in English, and jargon words are not always understood.

There is a tendency, too, to give only the barest information, and the consequences of misuse and abuse are not always described. Twenty years ago motor car operating and service manuals contained much more information and guidance than most of them do today.

For his part, from the minimal essential information which is supplied, the motorist is expected to ensure that his car is properly maintained. The Chrysler Company, which offers the 5/50,000 warranty, requires of any owners preparing a claim against the company that they shall provide documentary evidence, from an accredited dealer, that the car has been properly maintained throughout its life.

The designer will usually try to reduce dependence upon the user to a minimum. The effects of neglect of grease points have been mitigated by providing self-lubricating bearings in steering and suspension joints. This has had the additional advantage of increasing life and safety.

In other cases, conflicting needs prevent complete fool-proofing. We all know the motorist who reduces the life and safety of his tyres by careless parking against pavement kerbs, or by keeping them under- or over-inflated. There is a limit to the extent to which tyres can be strengthened to withstand such abuse without sacrificing performance and comfort. Hence, tyre satisfaction depends to some extent upon the user, who needs to be instructed. Yet how often is one supplied with an advisory booklet when one buys a new tyre?

Motor car batteries could provide another instance, and many more come to mind. In the case of ancillary equipment, documentation is required not only by the end-user but by the prime contractor as well.

> A few years ago a new type of popular motor car experienced considerable trouble through short-lived batteries. It was found that the battery had been installed close to the exhaust manifold. The battery manufacturer met the warranty claims, as well he ought to have done, because he had not provided the motor car designer with information regarding the maximum environmental temperature for satisfactory life.
>
> Here was a clear case of failure to define the capabilities of the product, and the contribution which the vehicle designer could make towards its satisfactory performance.

With so much electrical equipment being used in the household, here is a source of increasing hazard. Useful steps have been taken by indicating on equipment and domestic power plugs which colour of lead should be attached where. A great deal more could be done, however, by informative booklets warning against the dangers of short-circuits and of frayed cables, to mention only two sources of risk.

212

The technical manual is in fact a specification—of the scope and limitations of the product, of how it should be treated to give best performance, of what constitutes misuse and abuse. This important matter of written communication is relevant and has already been discussed.

Educating the User

The individual domestic user is educated by such information leaflets as exist, by the small amount of guidance which a label can carry, and by the occasional feature on radio or television. There would seem to be great scope for audio-visual education to be brought into the home, by means of these media. Educational campaigns would probably be no more expensive and far more effective than current forms of advertising. They could inspire confidence and gratitude in a way which is beyond the scope of most present-day advertisements.

When the customer is a large organization, which depends upon its equipment to earn its livelihood, there is a great financial incentive to obtain the maximum performance. Sophisticated equipment requires skilled operatives, and suppliers of such equipment take it upon themselves to train their customers' engineers and technicians. This practice can be traced to the early days of engineering, when the Boulton & Watt Company sent its engineers to install external-condensing steam beam-pumps at lead, tin, and coal mines, and to train local men to operate them. The practice was continued by British railway and textile engineers in many countries, and, later, by the early automobile companies, who trained the chauffeurs of their early clientele in the principles and practice of good driving and maintenance. It is practised today, notably by aircraft engine firms, manufacturers of computers, and electronic equipment companies selling defence systems to countries overseas.

Such firms set a fine example, albeit on a scale which is not appropriate to everyone. The aero engine divisions of the Rolls-Royce company, for example, maintain customer training facilities through which have passed upwards of 100,000 pilots, flight engineers, maintenance and repair engineers of their customers, military and civil alike, from more than 80 countries.

As has just been indicated, it is necessary to train users in more than the basic design of the product, the way in which it works, its scope and limitations, the way in which it should be operated to obtain optimum performance, reliability, and life. Certain routine inspection and maintenance operations may be required, and those which are to be carried out by the customer must be clearly defined, and the operatives must be trained in the necessary techniques. It can go almost without saying that routine maintenance must be kept to a minimum, since maintenance operations always entail the risk of 'finger trouble'. Hence, all maintenance schedules must be examined critically, and a continuous effort must be made to eliminate as many of them as possible.

The tutors in company training schools are exceptional men. They must have an intimate knowledge of the product, and of the user side of the business. They must also have the ability to communicate to people from many different countries, with different attitudes of mind, a great amount of practical information in a period which is usually 1 week, and seldom exceeds 2 weeks. Such training schools could, if their achievement were more widely recognized, provide the prototype practice for industrial and technological education generally.

Attention to Customer Complaints

Most of us will have experienced the feeling of disappointment and frustration on receiving the all too common meaningless response to a complaint of unsatisfactory product. The rare exception of an honest and helpful reply usually leads to a resolve to deal with that firm in future.

The 'complaints department' which is responsible for the despatch of the standard letter to the dissatisfied customer is usually acting in the belief that it is protecting its firm's interests. It behoves senior management to make sure that all employees understand that it is the customer who pays their wages and salaries. The better the customer is satisfied, the better their chances of long and rewarding employment.

Management should remember that the service engineers who visit customers are, even more than the salesmen who were the customers' first contacts, the real emissaries of the company. They should be instructed and trained accordingly.

> Showing how a customer could be left more satisfied after experiencing a trouble than he was before it occurred, W. Martin-Hurst described what had happened to an American airline operator. He was using British aircraft, with British engines which enjoyed a reputation for the highest quality. But the time came when something went wrong with these engines, and he had to call for help. Before he had fully realized that he was in trouble, swarms of engineers had arrived from England and the trouble was corrected immediately. Having seen how this matter was handled, the chief executive of the airline realized how lucky he had been. If he had not experienced this trouble, he would not have known the resources of service and maintenance which were available to support his operation.[3]

The Provision of Adequate Spare Parts

The management of a firm responsible for a new product should make an intelligent estimate of the components which might be prone to failure in use, and of their probable rates of failure. The estimate, which should be carried out in detail will require a joint effort by design, development, quality, and service engineers. It should be made as early as possible so that, ideally, spares can be made concurrently with production batches.

Provisioning of this kind is particularly important where the supplier has guaranteed the reliability of his products. The ability to maintain the equipment in use, by rapid replacement of defective components, is not only greatly appreciated by the user, it goes a long way too to reduce the financial liability of the supplier.

Obviously, it is vitally important that a careful balance should be struck between two opposing risks. Overprovisioning could cost a great deal of money, and lead to large stocks of parts which could become either unnecessary or obsolete. Underprovisioning could involve the customer in substantial losses, claims on the supplier for recompense, and a loss of confidence in the supplier. All of which goes to show how important to the continuing success of a company is the reliability of its products.

The large investment by airlines in modern aircraft, and the intense competition between airlines, is clear justification for the effort to maintain the availability of equipment at a maximum. The analysis which was made by Sir Ronald Holroyd, described on page 67, shows that in other fields of activity the failure to provide such minor but vital spares as seals for a pump can result in the loss of millions of pounds' worth of output. Engineers in manufacturing industry are inclined to feel that far better service should be expected of machine-tool builders. Stories are common of new machine-tools which have suffered an early failure and which have stood useless for many weeks because no spares have been available. The remedy is obvious.

Keeping the Job Moving: the Quick Fix

Experienced service engineers who have the interests of the customer at heart become expert in maintaining equipment in operation by 'quick-fix' remedies. Simple repair schemes, which must, of course, be approved by the engineering department, may be introduced. A current trouble may be minimized by changing the conditions of operation, or by introducing preventive maintenance. The life of parts which have proved to be inadequate can be limited, providing, of course, that spares are available, and the supplier's service engineers can help the user to cope

215

with the additional work load involved in the replacement of short lived parts.

Probably because of the high costs involved, it is the civil airline operators who are best able to provide detailed examples of the value of in-service restoration of deteriorated components, by relatively simple repair. Leamon and Pratt[4] have described how lack of repair schemes can, in the case of a large airline, lead to the accumulation of large stocks of parts awaiting reclamation, which can amount in value to more than $1m. They show too how the flame tube (US 'burner can') of a gas turbine engine, costing $1000, can be given a second useful life of 1800 hours by a simple inexpensive repair. As the airline processes 600 of these components per month, the savings are considerable.

Maintainability

The useful life of many types of product is limited by factors such as wear and/or corrosion, or failure to function due to accumulated dirt or products of combustion. The purpose of 'planned maintenance' is to extend the life of the affected components by corrective action, which takes the form of doing whatever is needed to restore the parts to full effectiveness. Thus, it may be necessary to clean parts which are becoming affected by corrosive by-products. Oil-ways may be obstructed by sludge, piston rings may become stuck by gummed oil, fuel burners obstructed by carbon. Oil in sump may become corrosive through contamination by condensed water, or by products of combustion, and risks of corrosion can be minimized by draining, cleaning, and by providing fresh oil. Electrical switches may have their life limited by deposits of copper due to arcing, or by burning of the contacts. Cleaning or replacement of small inexpensive parts can obviate breakdowns.

These activities represent maintenance at its simplest. With the increasing capital investment in modern civil aircraft, airline operators have been quick to realize the economic importance of maximum reliability at minimum cost. With the greatly increased cost and complexity of military aircraft, and the increasing reluctance of Governments to approve Defence Votes, air forces have been compelled to concentrate on achieving maximum operational availability of their equipment. Hence, during the past few years, maintainability has come to be regarded as a separate subject deserving of serious study. There is now at least one definitive text book on the subject.[5] It is accepted that the aerospace industry has become the pace setter in this field.

The best maintenance is no maintenance. This, however, requires action by design and development to increase the useful life of components and to eliminate the need for cleaning or lubrication. Bryson

216

et al[6] have pointed out that the prime factors leading to the need for preventive maintenance in the Fleet Air Arm are:

Defects.
Corrosion.
Component lives.
Modifications.
Electronics.
Spares.
Scheduled maintenance.

They advocated the study of maintenance at the design stage so that it could be made easier, the provision of easier access, and the development, among other things, of automatic checking equipment.

Great efforts have been made to reduce the amount of maintenance, to save down-time and labour costs, and to reduce the risk of 'finger trouble'. These objectives have been achieved by designing aircraft gas turbine engines so that the shorter-lived parts can be removed easily, with the minimum of disturbance to those parts which can be left to run on for a longer period. Inspection holes have been provided, so that optical probes can be used to determine when critical parts such as nozzle guide vanes have reached a stage where reconditioning or replacement is necessary. This lengthens the useful life of the parts while at the same time reducing the amount of maintenance required. United Air Lines follow policies known as TARAN ('test and replace as necessary') and RCOH ('reliability controlled overhaul'). The special characteristics of the aircraft gas turbine engine have been of help as facilitating:

(a) Ease of partial disassembly.
(b) Intensive inspection of risk areas by optical, ultrasonic, and X-ray methods.
(c) Monitoring of engine condition by analysing certain operational parameters (see Trend Analysis below).[7]

Vibration pick ups, with indications in the cockpit of the aircraft, can give warning of impending bearing failure, or of a broken blade. Magnetic plugs in the oil system can give an indication of incipient failure by collecting particles of steel. These methods enable the engine to be shut down and repaired, or removed for repair, before an expensive failure has taken place.[8]

Trend Analysis
In the next chapter, we shall see that normal behaviour in service conforms to certain definite patterns. Methods have been developed of

recording and analysing certain aerodynamic/thermodynamic parameters of the gas generator of the gas turbine engine, and mechanical parameters of the engine as a whole. Departures from the norm serve as danger signals.

Degradation of components affecting an engine's aerodynamic and thermodynamic performance can be detected by comparing the relationship of engine speed, fuel consumption, and exhaust gas temperature with the norm for a given air temperature and pressure at the engine inlet. Mechanical integrity can be confirmed by measuring vibration, the run-down time of the engine, oil consumption, and spectrographic analysis of the oil. These methods of 'trend analysis' have been used successfully by airlines, to indicate impending trouble. Engines have been removed in time to prevent failure and expensive damage. Other engines have been left untouched, to gain maximum value from them. Similarly the retraction and extension time of the undercarriage gear is a measure of its functioning. Brake drum temperature can be used to indicate impending loss of braking power.[9] [10]

The cost of the equipment required for this monitoring can be recovered from the savings due to one failure prevented. One airline has estimated savings in excess of $26,000 per aircraft per year, which with a fleet of 30 aircraft is a not inconsiderable sum. In addition, of course, there is the incalculable value of enhanced safety of life.[11]

Trend analysis might well offer comparable savings to operators of fleets of buses, trucks, and railway trains.

References

1. Eltis, E. *loc. cit.*
2. de Ferranti, Sebastian Z. 'Technology and Self-Determination.' Discourse at Royal Institution of Great Britain, 21 March 1969.
3. Martin-Hurst, W. F. F. The Managing Director, Rover Company Ltd. Proceedings of the Discussion on Quality and Reliability, Institution of Mechanical Engineers, 1963, London.
4. Leamon, R. J. F. and Pratt, J. T. 'Extended Engine Life through In-Service Development.' Society of Automotive Engineers, April 1966, New York.
5. Blanchard, B. S. and Lavery, E. E. *Maintainability*. McGraw-Hill Book Company Inc., New York, 1969.
6. Bryson, L. S., Illingworth, P. H. C., Thornton, F. S., Titford, D. G., Watson, P. A. L., Webber, R. H., Waller, E. W. de W., and Hammersley, E. J. 'Naval aircraft maintenance.' *Production Engineer,* London, September 1962.
7. Raymond, E. 'Product Support.' Lecture to Association of Aero Service Managers, March 1965, London.
8. Bowling, A. G. *loc. cit.*
9. Spraker, W. A., Loomis, T. P., and Bagly, F. C. 'Trend analysis as applied to gas turbine engines and aircraft mechanical equipment.' Proceedings Aero-

space Reliability and Maintainability Conference, May 1963, American Society Aeronautics and Astronautics, New York.

10. *Aircraft Engine Trend Analysis*. Two volumes. Battelle Memorial Institute, Columbus, Ohio, November, 1965.

11. Bowling, A. G. 'Lessons from engine experience.' Colloquium on Aircraft Reliability in Service. *J. Royal Aero. Soc.*, Vol. 70, No. 663, London, March, 1967.

16
Product Improvement

Few of the smaller firms in the nation's industrial complex possess the means to carry out enough market research, design, and development to ensure that the end-product will give adequate customer satisfaction. Many larger firms, who have the resources, and who ought to know better, regard the seller's market with a measure of cynicism. They are happy to try out a new product 'on the dog', depending upon market demand and the absence of competition to outweigh the long-term considerations of customer satisfaction and survival in markets which will some day be competitive.

Even companies which try conscientiously to evolve satisfactory products share with the less scrupulous ones the same tendency to neglect the best possible source of information regarding the success of their design enterprise—the way in which the product performs in the hands of the user. Yet information of great potential value is received by them from dissatisfied customers, whose letters are acknowledged, in many cases, by a reply which usually says that the trouble has never happened before, or that it is due to misuse or abuse by the complainant. It may be that suppliers who look upon their customers as their lawful prey do not deserve to benefit from this inexpensive and effective source of information.

> . . . no matter how good the design, how effective the development testing, and how good the manufacture and quality control, unanticipated troubles will be encountered when the product goes into field service. Realizing that it is proper performance in the field which is the overall objective—the pay-off to all the effort which has gone before—the wise industrial supplier tries hard to obtain early warning of troubles, to discover the reasons for those

troubles, and to eradicate them. Surprisingly though, there are still many firms which will spend a good deal upon development testing, yet fail to realize that here, in service experience, lies the best and cheapest source of information regarding the capabilities and behaviour of the product, in the precise and actual environment.[1]

Suppliers of sophisticated equipment, as in the aerospace industries, know from experience how difficult it is to estimate with certainty the conditions of use and environment with which their products must cope. This is largely because in such a progressive field each successive product represents a considerable advance in performance over its predecessor. Neither supplier nor customer can be entirely sure of the conditions which will be encountered. Also, when a new product reaches the user, new applications may be seen which will impose new demands. In consequence, the reliability which will have been strived for will often not be achieved at the outset. The efforts which are made by firms in the aerospace industries to collect information regarding the behaviour of their products result eventually in considerable improvements in reliability and value to the customer. Although the scale of the effort required in this field is beyond the reach of most other firms, the principles are so sound that they provide a classic example of good practice which can be followed with great benefit by firms of all sizes, in every kind of enterprise.

Despite the very considerable efforts which are made to evolve satisfactory aircraft, aircraft engines, and accessories, it is usually possible to improve reliability in service by up to 10 times, by learning from service experience.

Figure 16.1 shows the reliability pattern of a well-known American aircraft engine.[2] It is typical of aircraft engines generally, and it shows how the reliability is often at first disappointing. After about 50,000 engine-hours' of experience have been accumulated, it will have been possible to eliminate the more important causes of the inability of the engine to cope with the actual conditions of use. Engineers and managers in the aircraft industry appreciate the difficulty of foreseeing all the conditions of use in service. They realize that the causes of the high initial unreliability are, by and large, the actual conditions for which the product was really intended. There may have been some increase in the demand on the engine or the aircraft since it was first conceived some years previously. There may have been some misuse, but more often than not, and despite the great efforts which will have been made to forecast the actual conditions of use, only experience will evince the

221

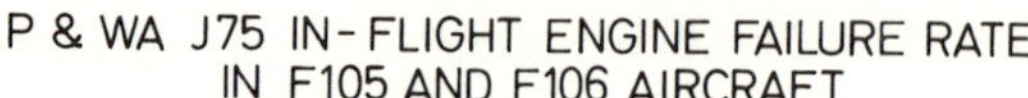

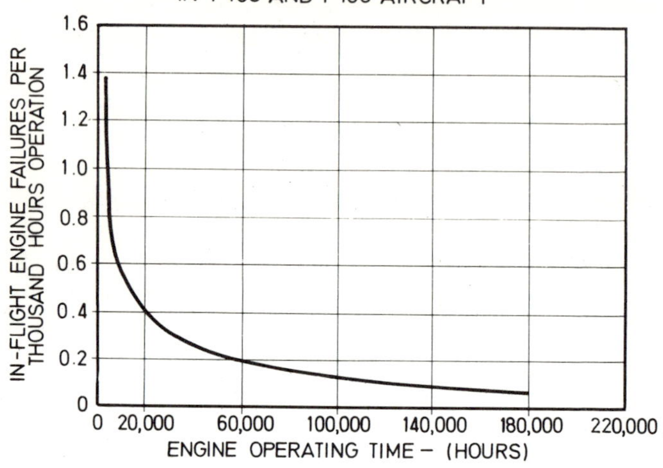

Fig. 16.1. Rate of Improvement in Reliability for an Aircraft Engine in Service.[2]

truth. A great capital investment will have been made by the customer—
an air force or an airline. It is vital that the product should be made
satisfactory.

This is done by collecting information regarding the way in which the
product is behaving in service, analysing the pattern of failures, weight-
ing troubles according to their importance, diagnosing their causes, and
eliminating these causes by modifying the conditions of use, or, more
likely, by design changes. These design changes must be proved by
development testing as being likely to be superior to the original design.
The customer must be convinced of their worth, they must be introduced
into service with least inconvenience to the user. And as early as possible
so that the costs of unreliability can be reduced without delay.

It may be thought that these conditions are unique to the aircraft
industry. The initial unreliability is due to the continual thrusting into
new levels of performance, but it must be remembered that great sums
of money will have been spent in the design and development phases.
In the more general case in other industries, far less will have been spent
on design and development, and although data is sparse there can be
little doubt that there is at least as much scope for reliability improvement through in-service development as there is in the cases just described. At the same time, in these other industries, the collection of data will
usually be a far simpler business than it is with, for example, aircraft
engines. There is all the less reason, therefore, why great advantage should

222

not be taken of this best and cheapest of all experience—of product performance at the hands of the user.

To recapitulate the necessary cycle of activites, there must be:

Information
(a) Collection of information.
(b) Analysis of service troubles.

Action
(c) Evolution of remedies.
(d) Introduction of remedies into service.

Data Collection

The action which a supplier needs to take to collect the necessary information can be seen to vary between wide extremes. At the level of consumer goods, of which most of us have personal experience, it is common to find that the least effort has been made. Yet it is here that it is relatively easy to obtain information of great potential value.

Even without encouragement, some dissatisfied customer will complain either to the manufacturer, his agent, or the retailer. All that is necessary, as a first approach, is that these complaints should be heeded, rather than brushed aside. They will comprise only a relatively small sample of the total of the troubles being experienced. The fact that they will have inspired unsolicited complaints will probably mean that they have caused some inconvenience, and they will certainly merit serious study.

At little cost, however, the amount of information can be increased appreciably. Customers can be told, either through advertisements or through dealers and retailers, that the manufacturer would welcome news about the product's performance in order to improve it.

Probably the most widely known example of effective involvement of the user is that provided by Marks and Spencer Ltd. Millions of people know that they can buy clothing and groceries in the sure knowledge that any complaint will be met by the immediate replacement of the unsatisfactory purchase, or by the return of the purchase price. All complaints can be identified by code marks. Reports of complaints are communicated by telex to the company headquarters, so that up-to-date knowledge of the cause and source of rejects is obtained. Immediate action can then be taken with the supplier concerned. The company is buying, at a low total cost, information which it puts to good use.

This market is a very large one, but the products are relatively simple. The motor car industry has fewer transactions, but a much more complex product, with a longer useful life than have the majority of items of clothing. The motor car owner is more concerned with the reliability of his purchase, which usually ranks second only to his house as a single item of expenditure. The customers of Marks and Spencer are more concerned with the quality of the purchase—its fit and finish, the security of its buttons and zip fastener, or the condition of the apples and bananas, at the time of buying.

Some motor car firms have set themselves the task of collecting as much information as possible, and putting this to the most effective use so as to achieve the maximum of reliability.

The pioneer in this field has been the Chrysler Corporation, which was the first to realize the value of reliability. In 1962, it decided to offer to its customers a warranty of 5 years or 50,000 miles. In less than 4 years the company's sales had doubled, and its earnings had increased considerably.

The customer is expected to provide evidence of correct servicing and maintenance, but it is vitally important to the company that it should have accurate knowledge of causes of failure so that it can reduce the warranty costs to a minimum.

It is necessary to maintain records on every car produced for a period of $5\frac{1}{2}$ years. This requires something like 8 million base records, each with three or more subsidiary historical records.

These records start with the initial history of the design status of each vehicle, and they are augmented by details of each warranty claim. Checks are made that the required maintenance has been carried out, the claims are settled, and the records are analysed by computer to show where corrective action is most needed.[3]

Within 2 years of instituting the scheme, warranty claims had been reduced to one third of their previous level.[4]

A less sophisticated approach enabled the Vauxhall Motor Company Ltd to make considerable improvements in the reliability of its vehicles. Delivery drivers and dealers were recruited to assist in the reporting back of defects discovered during delivery journeys and pre-sales servicing. Dealers were also helped to report back customer complaints, by being supplied with simple pro forma.

The Rover Company Ltd also bases its reliability improvement programme upon customer complaints, which are collected by the service department and investigated by the quality department. Trend curves are plotted to check that design changes have been effective. (Fig. 16.2.)

224

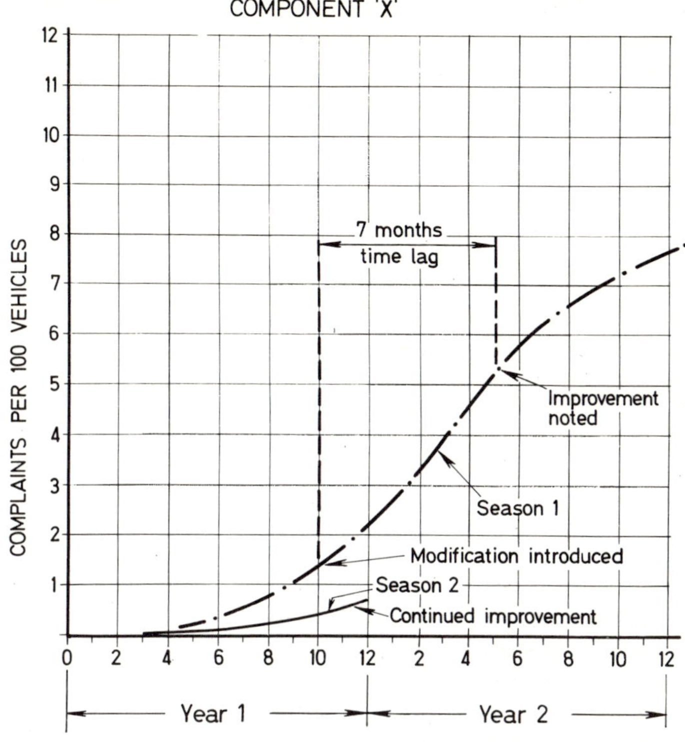

Fig. 16.2. Improvement due to Effective Design Change.[5]

Defects are classified into four categories, as involving:

(a) Safety of the vehicle.
(b) Company prestige (e.g. immobilization of the vehicle).
(c) High rectification cost.
(d) Customer irritation.[5]

With aircraft engines, the situation is still more complex. There will be several thousand engines of a given make in operation, with air forces and airlines in most countries in the world. Each engine contains two hundred or more critical and/or costly components. On their satisfactory performance and longevity depend the safety of aircraft and passengers, and the operating costs. The useful life of the engines will be of the order of 10 years with the first user, and perhaps as long again in the hands of a subsequent buyer of second-hand aircraft.

Leading aircraft engine companies have always known that it is essential to have accurate information regarding the performance and behaviour of engines and components. Formerly, the information was

225

obtained through the medium of defect reports, service engineers'
reports and defect summaries. With the great increase in civil aviation
and the much greater complexity and cost of engines, it has become
necessary to handle large volumes of information quickly and economically.

Rolls-Royce Ltd have developed a service recording and data
analysis system (SRDAS) which has been widely adopted by other
organizations in the aircraft industry.

Its purpose is to provide information to enable:

(a) Effective design changes to be made, to eliminate current
    service troubles.
(b) Future designs to benefit by being improved to avoid known
    troubles.
(c) Spare parts requirements to be forecast with greater accuracy.

The system depends upon the recording of data on simple,
specially designed cards which are completed either by the customer
or by the company's service engineers. These are returned to the
company's computer department, where they are processed.
Information is stored, and weekly, monthly, and random printouts
are made providing routine information and special analyses as
required.

Not the least valuable facility is the comparison of good performance with bad. For example, a given type of engine may
perform better in one type of aircraft than in another. The same
type of engine in the same type of aircraft may be more trouble-
some in the hands of one airline, or even one pilot, than in others.
The computer can help to detect such differences, leading to the
diagnosis of the cause and the rapid elimination of the trouble.

It is important to realize that no system is better than the information
which is fed into it from the field. The quality of the service engineers is,
therefore, of direct concern to management. Motor car companies
train their dealers. As we have seen, Rolls-Royce Ltd treat customer
training very seriously, but in addition they have always maintained a
strong and effective service department, with skilled engineers stationed
at company and customer bases across the world.

Inaccurate information can prove to be worse than none at all. A
company supplying engines for a military vehicle spent a considerable
effort investigating reports of overheating. It required a visit by a service
engineer to the station, to find that the radiators were clogged with mud.

Extracting Information from the Data

The information which has been collected must be put in order before it can be of use. When processed by computer, it is of course possible to extract information quickly, in any form. In all cases, the first action should be to identify the troubles and rank them in order of importance.

This will produce a Pareto distribution (see page 209). The vital few troubles, which are generally responsible for about 70 per cent of all the trouble, can be identified. Ranking the information in order of importance will make it possible to concentrate attention where it will be most effective.

> The chief executive of a company was recommended to identify the worst six causes of trouble in service, and to tackle those first. His reply was that there were no worst six, as a different trouble was reported to him every day. Investigation showed that in fact 50 per cent of all the trouble being experienced was due to a single cause. The majority of the daily reports were of troubles which were isolated ones, or which were few in number. Yet the chief executive punched his bell every morning and demanded a report on the latest, novel trouble just notified to him.

Figure 15.1 is taken from the report of faults found in motor cars in Sweden, during the compulsory annual examination. It indicates the typical Pareto distribution.

In the great majority of cases, once it is known which component or sub-assembly is causing trouble, it will not be difficult for the designer, the development engineer, or the production engineer to determine how to put things right. He can be aided, too, by the kind of testing described in chapter 9. The obvious essential step is to make sure that these people know that there is trouble, and that it is a matter of urgency to develop a cure.

Here, action by management is usually necessary to effect a proper system of communications, to ensure that responsibility is properly allocated, and that action is taken by the appropriate department. It is not at all unusual to find that the design, development, and production engineers will try to refuse responsibility for the basic fault, each blaming it on the other parties, or, more likely, upon the inspector.

It may be found useful to make one individual or department responsible for diagnosing the cause of trouble, and indicating which department should take action. Depending upon the size of the organization, the type of problem usually encountered, and the degree of collaboration and trust existing between departments, this responsibility might be taken by the chief executive himself, by the chief engineer or chief designer, by the quality engineer, or by the service engineer.

With more sophisticated and complex equipment which is faced with novel conditions, as in the electronics and aerospace and nuclear industries, diagnosis of cause may not be so easy. In complex equipment containing large numbers of similar items it can be important to know, as soon as possible, whether the first failure indicates a basic weakness of that component, or is merely a random, one-off incident due to a rare defect in quality.

Here, the pattern of failure, with time, can provide information of great value. It can enable the engineer to take early remedial action, and it can provide a sound basis for the planning of spares requirements. When a design modification has been introduced into service, it can provide evidence of the success or failure of the change. Figure 16.2, which is based on work by Witts,[5] illustrates this last point.

Types of Bathtub Curve

So far we have discussed product improvement in general terms. Since far more has been written about the reliability of electronic equipment than about mechanical products, an account of the great contribution made by electronic engineers is necessary.

Reliability was first recognized as a subject to be regarded as important in itself by the engineers concerned with the unsuccessful launchings of the early guided missiles. They evolved a statistical approach, which was entirely valid in the circumstances, where the end-product contains large numbers of similar components—diodes and triodes for example— which are vital to the satisfactory functioning of a complex system, yet whose integrity cannot be assured by non-destructive testing. Unfortunately, the impression gained ground that this was the whole of reliability. From this work was evolved the concept of the bathtub curve, so-called because of its shape.

It is important to understand the pattern of reliability behaviour, and to realize the different conditions which must be taken into account when considering failure patterns of mechanical equipment as distinct from electronic products.

Figure 16.3 shows at (a) a typical bathtub curve for electronic equipment, and at (b) a somewhat similar curve for an aircraft gas turbine engine.

In the case of the electronic equipment, the first part of the curve, which shows improving reliability, is sometimes known as the 'burn-in' period. It indicates a high initial unreliability, due mainly to the fact that with complex electronic equipment there is a high probability that

228

there will be a not unappreciable percentage of defective components. These will begin to fail as the equipment begins to be operational, and as failed components are replaced by new ones the reliability gradually improves. When all the unreliable components have been replaced, and when the unreliable components among the replacements have been replaced in their turn, the equipment comes to consist mainly of reliable components. There ensues a period of time during which the performance of the whole equipment is adversely affected only by random failures of individual components. These have been less than fully effective because of some atypical weakness which has been un-detectable, or due to some error of assembly. Later, as other functional components degenerate, through fatigue, wear, or perhaps creep, their useful life comes to an end, and unreliability begins to increase, as indicated by the rise in the curve, at the 'wear-out' period.

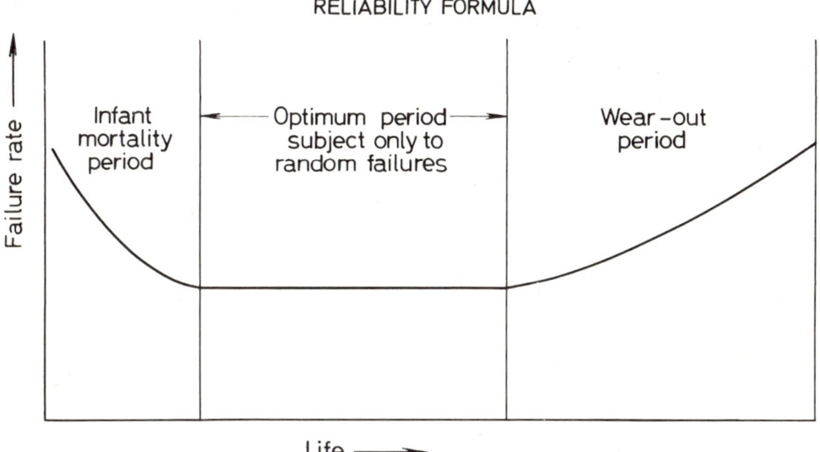

EXPONENTIAL FORMULA

$R = e^{-\lambda t}$;   Holds only for constant failure rates
R = Probability of survival for time t or longer
λ = Failure rate = $1 \div m$; m = mean time between failures
t = Time
e = Naperian base of logarithms = 2.718 +

Failure modes differ for different sections of Bathtub Curve (Analogous to human mortality) ; hence remedies also differ

Fig. 16 3a. The Bathtub Curve: The Classic Curve for Electronic Equipment.

The other diagram, Fig. 16.3(b), follows generally the same pattern. It is important to realize, however, that whereas in the electronic case the time-scale indicates the total life of each individual unit equipment, the time-scale for the mechanical equipment indicates the accumulating

229

life of a fleet of equipments—the engines of a given type in a given airline, say, or the engines in a fleet of buses or trucks.

The high unreliability in the early stages will be due, usually in small degree, to errors of manufacture or assembly, and to a greater extent to components whose designed capability proves to be inadequate for the actual conditions imposed upon them. With effective feedback of information, and as the learning phase progresses, the troubles due to errors of manufacture and quality control should diminish rapidly. New equipments entering service will, therefore, start with a higher initial reliability than the earlier ones. Meanwhile, the other troubles will have been reported back, action will have been set in motion to design, test, and produce modifications to obviate the causes of unreliability. As these are introduced into service, more and more equipments are freed of the causes of unreliability, and the performance improves until all equipments entering service give maximum reliability from the beginning of their working life.

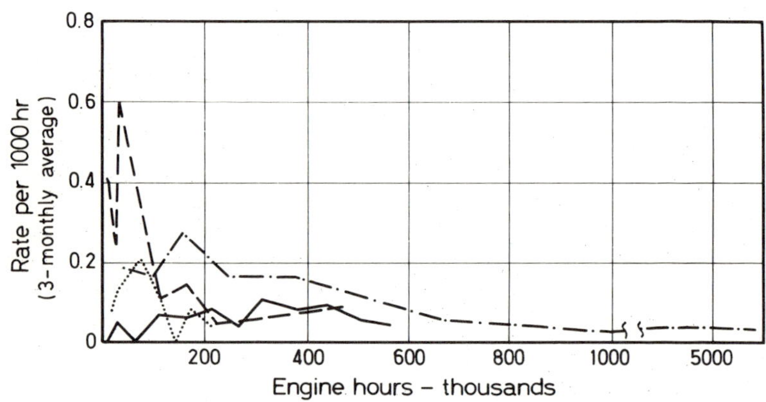

Fig. 16.3b. The Bathtub Curve: Typical in-flight Shutdowns for Civil Aircraft Engines.[9]

The first part of the curve, therefore, represents an improving situation, in the product type as a whole, as causes of unreliability are detected, understood, and remedied. This contrasts with the unchanging pattern of the typical electronic equipment.

It is interesting to note that in the case of the American military aircraft on which Fig. 16.1 is based, optimum reliability is not achieved until about 100,000 engine-hours of experience have been accumulated. Bowling has shown that in the case of civil aircraft, where maximum effort is applied to reduce financial losses, optimum reliability is achieved after about $2\frac{1}{2}$ years' operation. Motorists who suffer from teething troubles in new models will be interested in this.

230

The importance of the patterns of reliability which we have been discussing lies in the information which can be obtained regarding the probable trend of events. This in turn can be of value by indicating the best action to take. For example, if it were to be assumed that the initially high level of unreliability was unchangeable, greatly excessive stocks of spare parts would be ordered. On the other hand, an increase in unreliability above the minimum which will, hopefully, have been achieved should be taken as a danger signal, indicating that some new cause of unreliability has arisen. If adequate spare parts are not available, serious disruption of the operation may follow.

The foregoing arguments apply to the overall picture—the reliability performance of the equipment as a whole. The curve of failures with time, of an individual component, can help considerably in diagnosing causes of failure, where these are not readily apparent. It can help us, too, to be assured, with a minimum of testing, that a design change is in fact better than the original part which it replaces.

This is achieved by means of a technique evolved 20 years ago by Weibull,[6] but only recently applied to problems of reliability. Its practical application has been urged by Shainin[7] and others. As with all new techniques, however, there is always the danger that they will be adopted uncritically and overenthusiastically. A valuable service has been rendered by the late Bompas-Smith,[8] who made a thorough investigation of the scope and limitations of the Weibull method by applying it to historic data from well-documented records of service troubles. He concluded that the method was most effective when blended with engineering judgement. This advice could with advantage be applied to most other techniques and technologies.

It is *not* necessary that management should take the trouble to understand the basic theory of the Weibull distribution. It *should* be realized that here is, readily available, a method of achieving reliability with a minimum of testing.

Weibull is not a statistician, and his distribution is an empirical one. He had to understand statistics, but it was his knowledge of the needs of the research and development engineer which has led to a work of great value.

Remedial Action

Information without action is valueless. There are times when service engineers who have provided adequate data despair of seeing recurring faults remedied. Even when responsibility has been accepted, the designers who have worked upon the particular product will be heavily involved in current projects. They will find it difficult to accord the same priority to a design which by now will be several years old. Even when a design change has been made and proved, there remains the difficulty of persuading the user of its worth. The cost of modifying a fleet of buses, trucks, hire cars, or aircraft, can be high, and it must be offset against the estimated savings. Assuming that the modification will be successful, the savings will be equivalent to the costs of the current trouble, which might include lost revenue due to non-availability of the product, less the value of the spare parts in stock, and less the cost of spare parts of the new design. In one case, an airline had lost £250,000 in secondary damage, due to an engine bearing failure. The modification cost £35,000, across the fleet. The timely adoption of the modification terminated a running loss which could have doubled the £250,000.[9]

Military users are often less quick to accept modifications. The officers involved in the decision are on short term appointments, and have probably not been concerned with the trouble for long. The decision might involve the scrapping of expensive stocks of spares, distributed widely across the world. There is usually some lingering doubt about the probable success of the modification. Bowling, however, suggests that it is more expensive not to buy a modification than to have to pay for two attempts to remedy a service trouble.

References

1. Nixon, F. 'Whose job?—responsibility for reliability—the manufacturer's viewpoint.' Lecture delivered at Royal Military College of Science, Shrivenham, July, 1967.
2. Schmickrath, B. A. 'Material Reliability Applied to the Design and Development of the TF-30 Engine.' 6th Navy-Industry Conference on Material Reliability, Washington D.C., November, 1962.
3. Redmond, G. H. and Webster, F. M. 'Chrysler's 5 and 50 warranty program.' Automotive Engineering Congress, S.A.E., New York, January, 1966.
4. Osann, F. 'Commercial reliability—the Chrysler 5 and 50 warranty.' 11th National Symposium on Reliability and Quality Control, AIEE, ASME, ASQC, IRE, Miami Beach, Florida, January, 1965.
5. Witts, M. T. 'Problems associated with the production of a quality car.' Lecture to Coventry Branch, Institute of Engineering Inspection, October, 1965.
6. Weibull, Waloddi, 'A statistical distribution function of wide applicability.' *J. Appl. Mechanics,* Vol. 18 No. 3, September, 1951.
7. Shainin, Dorian. Seminars in Britain and the United States, 1968–69.
8. Bompas-Smith, J. H. 'The determination of distributions that describe the failures of mechanical components.' 8th Reliability and Maintainability Conference, Denver, Colorado, July, 1969.
9. Bowling, A. G. 'Lessons from engine experience.' Colloquium on Aircraft Reliability in Service. *J. Royal Aero. Soc.,* Vol. 70, No. 663, London, March, 1967.

PART 6

For Action by Management

17

Getting Things Going

A wide experience of firms large and small, in many countries, with products varying from nuclear submarines to jerricans, has shown that there are some in every sphere whose quality and reliability practices are well above average. Invariably, these have been companies where the initiative has been taken by top management.

This is confirmation of the point which has been stressed repeatedly in earlier chapters, that the prime responsibility for effective action devolves, inexorably and inescapably, upon senior management. Where this has been appreciated, the fact of authority being derived from the senior executive, and the broad viewpoint which he brings to bear, have meant that satisfactory results have been achieved quickly.

In some firms, the management has been deterred from taking a similar course because of a suspicion that something more was necessary than quality control in its narrow sense. They have been reluctant to allow an upheaval in their organization which might be a flash-in-the-pan, with no lasting beneficial result.

It will be shown how management can convince itself of the value of the Q & R approach by first achieving a success in a limited area. When this has been achieved, the way will be open for management to take the ultimate step, of involving every individual concerned in the objectives of the enterprise.

Making a Start

Money talks, and an effective plan has been first to identify and tackle a problem which offers a reasonable possibility of being solved fairly quickly and which will result in clear savings. This will convince senior and middle management that there is something in this quality business, and that further action will be justified. It will enable senior management to establish a lead, by demonstrating its understanding of the basic principles involved in the new approach.

Breaking Through to Higher Levels of Performance

The value of this approach has been emphasized by Juran. He has evolved a diagram from which Fig. 17.1 is derived, which illustrates his argument.

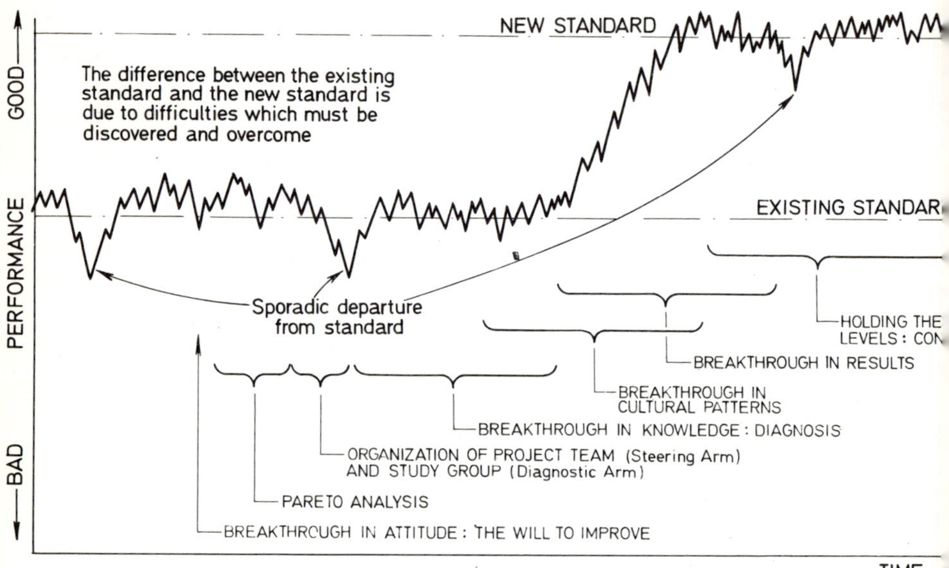

Fig. 17.1. Breakthrough to Higher Levels of Performance.[1]

It is a characteristic of most people to want to live comfortably with the *status quo*. It may have come to be taken for granted that a scrap rate of 5 per cent, that is, a performance rating of 95 per cent, is the norm for a particular process. A situation may arise which reduces this performance. When this is recognized, all resources are mobilized to 'fight the fire'. When the trouble has been identified and eliminated, or when, as often occurs, it has disappeared of its own accord, everyone is happy. The effort is relaxed, and routine resumes its usual tempo.

Why, asks Juran, should this state of affairs be accepted as normal? Why should not the effort be made to improve the average performance, to say 97 or 98 per cent? This is the same as tackling and eliminating or

236

diminishing a chronic cause of scrap, of which so many successful instances were reported during Q & R Year in this country.

A high proportion of managers simply have no time for Breakthrough because they cannot leave the treadmill of Control. They do *not* need to be convinced of the merits of Breakthrough. They *do* need respite from the never-ending emergencies and crises . . . they require help from upper management, with an assist (*sic*) from the staff people. The harassed manager can still find time for participating in the *guidance* of Breakthrough. But the remaining parts of the sequence must be worked out by other people.

The smaller company is inherently more sensitive to the need for change, more nimble in responding to the need. The big company must, in due course, break itself up into decentralised operating units, and the need to respond to change is one of the main reasons.[1]

(Parenthetically, it can be stated that Juran showed prescience when he wrote the foregoing. Some of the largest industrial organizations have already set the pattern by breaking up their organization into smaller, more manageable units.)

Management Initiative

The first and essential requirement is that senior management should be resolved to make a concerted effort to achieve a worthwhile improvement in some area or other of the organization.

The first action must be to bring together the heads of all the departments concerned, to form an action committee. The composition of this committee should be determined with longer-term developments in mind. Ideally, it should comprise responsible representatives of design, development, production engineering, manufacturing, purchasing, inspection, service engineering, and sales. The senior executive who will have set up this committee will take the chair. It will be his task to communicate to the committee as a whole his enthusiastic resolve that an improvement in performance can and shall be made, and that this will be demonstrated by tackling a limited objective first. It will be especially important to make every member of the team feel that he shares in the collective responsibility for the performance of the organization as a whole.

This may not be easy, because there may exist deeply entrenched departmental attitudes, which tend to be isolationist and inward-looking. The members of the committee might at first be at some pains to prove that everything is someone else's fault.

Here is a challenge to the senior executive. He should, however, with the help of the arguments and evidence produced earlier in this work, be able to convince the members of the committee that no department can work in isolation from the others, that each reacts upon the department either side of it, and often beyond that. Each of them must, therefore, bear some share of the responsibility for any shortfalls in the efficiency of the organization as a whole. It should be laid down as a basic principle that whatever the outcome, no blame will be attached to anyone. The intention is to achieve improvements which will benefit the company, not any individual department.

Having generated a measure of communal feeling, the members of the committee should be encouraged to express their views on what they consider to be the most serious problem confronting the organization. In a predominantly manufacturing unit it might be high rates of scrap. In another, customer complaints might be leading to reduced sales, because of unreliability, poor deliveries, or high prices. A competitor might be offering a product at a lower price, or one having higher performance. The quality and delivery rates from vendors might be a source of embarrassment, or labour relations might have deteriorated in some part of the organization.

On rare occasions the committee may be unanimous regarding the major trouble, in which case the course is open for action. More usually, there will be divergent views. It is the duty of the chairman to reconcile these as far as possible, and then to call for opinions on the relative importance of the troubles. All views should be recorded, and ranked in order of aggregate importance. In other cases, as when causes of spoilt work are the subject of inquiry, there will probably be found to be many opinions, but no fact.

The determination of fact will necessitate a deliberate effort. The absence of the data which the committee will have indicated as being required will usually reveal that no individual or department is responsible for collecting and reporting the facts. As a rule everybody will have been too busy 'fire-fighting'. Here is the first indication of the need for an additional facility which may eventually take shape as a quality department. For the time being, the chairman must undertake to provide the services of an individual to seek out the missing data. What kind of man is chosen will depend upon the problem area, but often it will be found that a keen young trainee, seconded to the department concerned, initially for the sole purpose of eliciting information, will be all that is needed.

More often than not, once the trouble has been identified, the remedy will be obvious. In other cases, the causes may have their roots in more

238

than one department, and the remedies may be more obscure. In such cases, diagnosis can sometimes be achieved in committee, but more usually it will be necessary to allocate an individual or a group to study the problem in greater detail.

When a solution has been identified, it is not improbable that more than one department will be required to make radical changes. These departments may have to sacrifice some of their preconceptions, and here again the tact and leadership of the chairman will be called into play. It might, for instance, be necessary for the designer to increase a dimensional tolerance, to change the design of a detail component to facilitate manufacture, or to eliminate a cause of failure. It will be important to get him to do this without losing face. In other cases, concessions by more than one party may be required.

The Simpler Case Worthwhile improvements have been made, quite easily, in many cases. It is often found that good and able departmental heads, once their interest has been aroused, will not only be able to identify the main sources of trouble; they will also see what is causing these troubles, and what should be done to eliminate them. Surprising facts have often come to light. There has been a lack of tools for the job, or a lack of drawings, or undecipherable instructions, or no instructions at all. This last has been found to be a cause of trouble even in the most highly organized and critical operations. How much more likely is it to arise in less formal enterprises! A machine or an operation may be consistently working out of specification.

More often than not the individual directly concerned, usually the foreman, will have become convinced that it is useless to complain, because it has been his experience that action is never taken. He will have come to accept a shortfall of 5 or 7 per cent of output, due to scrap, as inevitable, and apparently acceptable to 'the management'. He will be delighted, and his cooperation will be assured, when he is convinced that top management really means business.

> The present writer was approached by a local company, which stated that they had failed to identify quality costs as great as had been postulated. A visit was paid to the plant, and it was found that there was in fact little scrap, as most of the incorrect products could be rectified.
>
> No one had thought to assess the costs of rectification. It was found that records did in fact exist, and that they were easily available because a job card was raised, which specified the 'allowed hours' for each rework incident. It was a simple matter to tot up the total of rectification work for each shop, and to present it to

the foreman concerned. The effect was immediate and striking. 'We know how to fix that trouble', said the foreman. 'We hadn't thought it mattered, as nobody had ever mentioned it'—an innocent but searing comment on higher management.'

The More Complex Situation In more complex situations, Juran has given valuable advice to the senior manager who is determined that a breakthrough to higher levels of performance shall be made. He has listed the major tasks as being:

1. To bring about a change of attitude, i.e. to get departmental heads to realize that improvements are possible by their working together.
2. To identify causes of trouble, and to rank them in order of importance—the Pareto principle.
3. To seek out the information necessary to enable causes of trouble to be identified.
4. To prepare recommendations for dealing with the troubles.
5. To agree the necessary actions and to follow them up to ensure that they are put into effect.[2]

Some useful guidance of a detailed nature is given by Juran in his book *Managerial Breakthrough*.

The success which will almost invariably result from the initial effort will arouse a desire for more. At this stage, it will be timely for senior management to give serious consideration to the setting up of a quality activity, along lines described earlier. Having succeeded in improving relations between departments, the climate will be appropriate, and the new department will have a far higher chance of success than would have been the case had it been injected into a resisting organization of highly individual departments.

Maintaining Cooperative Effort

Having achieved a measure of cooperation between departments, management should ensure that this is maintained and developed. The interest of the departments will have been aroused, but positive action will be needed to establish the cooperation on a permanent basis. This requires a formal plan, which must be established by senior management.

An example of good management practice is provided by EMI (Electronics) Ltd. The effectiveness of the quality organization is upheld by a pyramid of committees reporting ultimately to the managing director of the company. It is clearly recognized that each technical activity bears

240

its share of responsibility towards the quality and reliability of the end-product. This is reflected in the composition of each committee, as shown below.

1. *Quality Policy Committee*
Chairman: The managing director.
Members: Director of research and engineering, director of production, quality assurance executive (secretary).
Purpose: To formulate company policy on all matters affecting quality of design and quality of production. Meets three or four times annually.

2. *Quality Executive Committee*
Chairman: The quality assurance executive.
Members: Engineering manager, chief scientist, buyer, production managers, quality manager (secretary).
Purpose: To lay down the procedures necessary to implement the quality policy of the company: to direct corrective action when required: to present situation and progress reports to the quality policy committee. Meets quarterly.

3. *Location Quality Managers' Policy Committee*
Chairman: The quality assurance executive.
Members: The quality manager of each factory or work location.
Purpose: To coordinate quality assurance policies and practices throughout the company. Meets quarterly.

4. *Location Quality Committees*
Each work location has its own committee.
Chairman: The location quality manager.
Members: The manager of the unit, the manufacturing manager, the buyer, the manager and the quality assurance engineer of the product division, the location quality manager (secretary).
Purpose: To deal with quality problems in the area.

By this means each activity of the company is involved in the achievement of quality, and each committee is ultimately accountable to the chief executive.

Keeping Things Going The system just described provides an excellent example of the resolve of a company's directorate to ensure continuity of purpose. The benefits of the Q & R approach, of the close cooperation of departments, and the welding of them into an effective total effort are permanent and continuous. They are not something which is achieved just once, as the result of a single concentrated effort.

Every day some incident will arise, every so often new equipment or process will be introduced, some new design will be adopted, and all of these will introduce problems which, if not tackled along the lines described, will result in further inefficiencies.

It is management's responsibility to ensure that there are no friction losses at the interfaces between different departments. This requires that all those concerned should clearly understand what is expected of them, which in turn means that there must be an adequate and effective system of communications. The EMI plan ensures that the chief executive is in touch, through the committee structure, with each activity affecting the overall objective of customer satisfaction. Moreover, he is seen to be interested and concerned.

Too often, the setting up of a quality department, without integrating it into the organization on lines such as these, has one of two results. Other departments either leave it to 'take care of quality', or through sheer frustration the quality department tries to insist that other departments should do their job to the satisfaction of the quality department.

This last is a dangerous possibility. The literal interpretation of the Ministry of Technology's Specification AvP.92 ('Specification of Quality Management Requirements', June 1965), could lead to this very thing. It could guarantee that there would be friction, strife, and non-cooperation in the organization.

For the benefits already gained to continue, management *must* play its part, to ensure continuing cooperation and the deployment of all the different departments towards the desired end.

References

1. Juran, J. M. *Managerial Breakthrough*. McGraw-Hill Book Company, 1964.
2. Juran, *loc. cit.*

18

The Involvement of People

Having succeeded in bringing together into an effective working team the relevant departments of the organization, management will have demonstrated its ability to achieve its more obvious but not always recognized responsibility. Its final step, to achieve the ultimate payoff of the Q & R approach, must bring about the total involvement of all the people who are concerned, directly and indirectly, in the objectives of the enterprise.

We have seen how extensive is the complex of companies and services, and their employees, which is involved in the success of a single producer of an end product. In each individual case, 'the management', 'the firm', 'Smiths', or 'Browns', by whatever name the employer is identified by employees, is in fact a go-between, bringing together a number of people into an organization for the purpose of supplying goods or services to its customers, to the profit of all those who support its enterprise. At one or more removes, employees are suppliers to the market from which their wages and salaries are ultimately derived. In their turn, they are customers, in a big way when we consider people in the aggregate, of many more suppliers of goods and services.

Many managers are conscious of the importance of the support which is necessary for their employees, in the way of housing, transport, public services, retail shops and so on. The study of the total system is important to the efficiency of the individual company, and it is the concern of chambers of commerce, local and regional and national government. Managements generally, however, fail to realize the advantages which can be obtained by looking upon their employees, and their vendors and suppliers, as an important part of their own 'system'.

National Quality and Reliability Year, 1966–67, helped a number of firms to realize this. For maximum overall efficiency, it should be self-evident that management should avail itself of the total aggregate of skills and abilities that are possessed by its employees and by the employees of its suppliers. For this to be achieved, it is necessary that every individual should be looking, and pulling, in the same direction.

Pointing the Direction
Management's first action should be to inform its employees and the supporting firms of its basic policy.

When a company is new, and starting with a clean sheet, this is relatively easy. When a company has been in existence for a time long enough for attitudes to become established, things are more difficult, but a firm declaration of policy is all the more important. That it is never too late is shown by the Harvard School of Business Administration, which had been in existence for 50 years before its classic statement was evolved (see page 4).

A few examples will serve as a guide:

The Honda Motor Co. Ltd, Japan

> *Company Principle* Maintaining an international viewpoint, we are dedicated to supply products of the highest efficiency yet reasonable price for world-wide customer satisfaction.

This declaration of policy is displayed boldly in the factories and offices.

Metalurgica de Santa Ana SA, Spain

The quality manual of this company opens with the following paragraph:

> 1. *Quality Policy of M.S.A.*
>   The quality problem affects the whole Company.
>   The Quality of today is the guarantee of tomorrow.
>   For improving our position with regard to the competition and to increase the benefits and reputation of our company it is necessary to make products with sufficient quality as to satisfy the requirements of the market.
>   The Board of Directors of M.S.A. has adopted the policy of putting on the market only the products having sufficient quality as to deserve the customer's satisfaction during the effective life and during the functions for which they have been designed.

(It should be noted that both companies quoted above clearly recognize the importance of reliability.)

244

Mitsubishi Heavy Industries Ltd, Japan

'Quality first, no compromise on quality', is the traditional policy of the company.

Rolls-Royce Ltd

It is the policy of Rolls-Royce to maintain a level of quality and reliability of its products and services which will be recognized by our customers as being surpassed by none of our competitors.

Sears Roebuck and Co., USA

Customer satisfaction is an essential to Sears' long-range success. One important requirement is that products will be of high quality. A product is of high quality only when it meets all the reasonable expectations that our customers have set for it, considering its price.

Reasons should be given supporting the decision to start a drive for increased efficiency by initiating a Q & R policy. It is important that this should be done whether or not a new quality and/or reliability department is to be set up. The reasons will vary according to circumstances. In all cases, the importance of increasing the competitiveness of the company's products should be stressed. An immediate cause for the decision could be the success which had attended the first experiment, and the desire to obtain still greater savings. There might be a decline in the company's share of the market, due to increased competition, or reduced demand. There might have been a policy decision to compete for the business of a nationalized industry such as the Central Electricity Generating Board, which imposes contractual quality requirements on its suppliers. An existing customer might have issued new quality assurance requirements, as in the case of the US Department of Defense, our own Ministry of Defence, or the Ford Motor Company and other large purchasing organizations. Legislation might have been introduced requiring a new look at the company's Q & R performance if existing markets are to remain open, e.g. as by the US National Traffic and Motor Vehicles Safety Act, 1966, or the British Trade Descriptions Act, 1968.

In some cases, significantly in Japan, where the one-union one-company practice and the traditional respect for authority of the Japanese worker allows management to adopt a paternalistic attitude, the announcement of company policy is supported by exhortatory instruction.

The Honda Motor Co. Ltd promulgates the following:

Management Policy
(a) Proceed always with ambition and youthfulness.
(b) Respect sound theory, develop fresh ideas and make the most effective use of time.
(c) Enjoy your work and always brighten your working atmosphere.
(d) Strive constantly for a harmonious flow of work.
(e) Be ever mindful of the value of research and endeavour.

The Aircraft Division of Mitsubishi Heavy Industries Ltd has both a policy, and a motto. The latter is, more strictly, a pattern of activity, and the first a slogan.

Quality Assurance Policies
'Quality first, no compromise on quality' is the traditional policy of the company.
It is our motto to

1. Bring the entire organization into activities.
2. Research and improve steadily.
3. Take thorough preventive action.
4. Work and inspect on an assured basis.
   to manufacture reliable products.

The important thing to appreciate is that the message reaches every employee.

Arousing an Interest
Whatever may be the overall plan of campaign, management will help to ensure its success by seeing that two complementary approaches are followed. Both of these derive directly from Q & R. They are based upon the belief that the great majority of people would rather do a good job than a bad one, and that they would far rather be interested in what they are doing than not. The first step should be to identify employees with the product, the second to let them learn something about the customers who enable them to stay in employment.

Identifying the Employee with the Product An organization whose employees identify themselves with the company itself can consider itself to be very fortunate indeed. When the firm happens to be a long established one, it is not always appreciated how much is owed to the managers of earlier generations, who worked hard to establish a tradition of job and customer satisfaction. It is important that today's manage-ment of such companies should realize the value of such a tradition, and work hard to maintain it. The Q & R approach is really no more than

246

that which was commonly followed by the founders of the fine traditions of the older British firms, modified and developed to accord with modern jargon and technology.

The first step should be to inform the employees of the company's background, why it aims to satisfy a particular market, and what are its main products. This is best done by means of a company journal, supplemented by displays showing the complete end-products and their constituent parts. It is often forgotten that important people, those who build the products, seldom see the lavish displays which are mounted at trade fairs.

Types of house journal vary from the chatty newspaper which contains little beyond personal news of employees—their marriage, retirement, or demise, with reports of social and sporting activities—to the more informative ones about the company and its products. Some deal with both areas successfully, but many ignore the second one. Some firms, as a result of Q & R Year, have supplemented their social newspaper with a bulletin appearing three or four times a year, which is solidly but interestingly informative about the company's products, about new design features, and about new processes of manufacture. Publications of this kind have been enthusiastically received.

An employee likes to know what contribution is made to the end-product by the work which he does. A feeling of responsibility is engendered when it can be seen that their component has an important part to play.

> During Quality and Reliability Year, employees in the Lucas organization were encouraged to visit work stations upstream and downstream of their own. They saw the finished product, a motor car lamp say, of which the pressing made by them was a minor but vital component. They saw the finished product undergoing tests, and for the first time realized why the stipulated accuracy was important, to ensure that the lamp should be weathertight.

Nor should the employees of sub-contractors and vendors be forgotten.

> During the Second World War, enthusiasm was generated and output increased in the works of small sub-contractors, by even simpler means. A large coloured poster of a well-known bomber aircraft was supplied to each works. An arrow indicated the approximate position in the nacelle where the inconspicuous and humble but necessary products of the employees' work were fitted. The girls were now identified with the product. They realized that

the washers made by them had helped in last night's bomber raid. They now knew they were making a worthwhile contribution, their feelings of satisfaction and responsibility were greatly enhanced, and output rose.

In companies concerned with products the integrity of which is required to be high, for example aircraft engines or motor cars, another method has been successful. Short lectures, illustrated by slides or film, have been given, largely in the workers' lunchtime break. The way in which the end-product works has been described. This has been followed by an account of the particular function of the parts made in the particular shop. The strength and safety requirements have been described, and the operators have realized for the first time the great importance of their work. Knowing how much depends upon them has had a tonic effect. Sir Denning Pearson has said:

> In my opinion, Quality and Reliability—Q & R—provides management with one of the best ever means for involving people in the total enterprise.[1]

Identifying the Employee with the Customer It is but a step from the product to the customer. The interest aroused, by the employee being helped to understand that he makes a distinctive contribution to the product, is enhanced and amplified if he can be made to realize that he owes a responsibility to his customer and that the product of his work is the means of doing this. An attempt should be made, therefore, to let the employee meet representative customers, if possible on their own ground. Where this is not feasible, customers might be invited to visit the works and address the employees.

> During the Second World War, young and highly embarrassed pilots of bomber and fighter aircraft were given a change, though not a rest, from their normal duties by being sent to give talks to hordes of adoring girls who had helped to build their aircraft.

As part of the Q & R Year activity, firms in the aircraft industry sent groups of employees, sometimes chosen by lot, sometimes as a reward for performance, to visit Royal Air Force stations and to see at first hand something of the problems of the users of the equipment that they had made. Other firms arranged for selected employees, again chosen by lot, or as a reward, to be given flights in aircraft powered by engines built by themselves.

> Marks and Spencer Ltd encourage their suppliers to select a small number of their employees, winners of a competition perhaps, to be their guests for a day in London.

On arrival they are welcomed and invited to see the latest products of their machines on display for the first time, at one of the leading stores. They hear the comments of customers, and report these back to their colleagues.

Planned Motivation

Another approach to the task of arousing the interest of employees in the aims of their employer has become popular during the past few years. The key word is 'motivation'. Practically concurrently, in Britain, the United States, and Japan it has become fashionable to mount campaigns with the intention of 'motivating', converting, or convincing employees of the importance of product quality, and recruiting them to its support. These campaigns vary only in the amount of ingenuity which is applied, to give them a measure of novelty and, by implication, appeal and potential success. Not unexpectedly, promoters of schemes which have been successful in one country feel that they should be equally successful elsewhere. Ignoring the existence of the different national and local characteristics which prevail in other countries, they urge their pet scheme as the universal panacea.

This is not to deprecate the value of promotional schemes which are appropriate to the circumstances. Management needs only to be warned not to accept proposed schemes uncritically. They should consider all the circumstances, the personality of their own organization, and the probable reactions of their employees to the style and the character of the schemes, before adopting any particular one.

It will be interesting and illuminating to examine the different schemes which have had some success in different firms in different countries.

Campaigns Most of the schemes which have so far been adopted have been promoted as a 'campaign'. The intention is to gain the attention of everyone in the organization, and focus that attention upon a particular subject, or upon a facet of a subject, for a sufficient period of time to achieve an impact and, hopefully, a lasting effect.

In Britain, campaigns have usually been mounted for a period of one year. We have had National Productivity Year, followed by Quality and Reliability Year. Japan, too, has had a Quality Year. In both countries, large numbers of firms mounted campaigns concurrently, and of like duration.

There is value in limiting the duration of a campaign. People can tolerate only a certain amount of concentrated exhortation, and a year is probably a little more than long enough for a high tempo effort to be maintained. The pressure upon the organizers must also be borne in mind. A successful campaign needs careful planning, and many months of preparation are usually required before a campaign can be launched.

On the other hand, and perhaps indicative of great stamina, the Zero Defects campaign in the United States has already run for several years, and although it has somewhat modified its arguments, it is still going strong.

In planning a campaign, it is important to ensure that the right subject has been chosen. Once launched, it will not be possible to change the title or the basic idea. Full consideration should be given, too, to the long-term implications. This will take time, and perhaps the best advice to anyone about to mount a campaign is not to start it too quickly. Indeed, the Lucas organization, which has always been a supporter of the Q & R concept, ran a campaign for a trial period of a year, two years before Q & R Year. The purpose was to learn from experience, to help other firms by passing on this experience, and so to ensure that the Q & R Year (1966 to 1967) would be the success that it was.

Ideally, a campaign should be planned so that it can be both a springboard and the basis for a continuing policy. In many cases, it will serve as the necessary initial indoctrination period, which will enable effective and acceptable policies to be established.

A report which was published by the British Productivity Council[2] gives details of the campaigns which were mounted by companies in a variety of industries, and these provide a useful guide.

Open Days An activity which has been found successful as a means of arousing interest in the company has been the holding of open days. These have been an annual feature for a long time in some firms. Many more were held in support of Q & R Year. For an employee to be able to take his family into the company's premises and to show them the environment in which he works, details of his own job, and of those of his friends, and something of the end-product, provides a boost to his own morale.

The most successful of these events have been those in the preparation of which the employees themselves have participated. This has aroused a spirit of competition between departments, and pride in achievement, which is lacking when the displays are merely those which have been prepared by a professional organization for use at a trade fair.

The mounting of an open day can cost as much or as little as a company is prepared to spend upon it. It should be realized that success depends far more upon the enthusiasm and cooperation of the employees than it does upon the money expended. In fact, lavish displays can have an effect opposite to that intended. Since the company is acting as host to

250

the employees and their families, it is to be expected that light refreshment will be provided. The additional expenditure will amount to no more than that required to provide paint and transport, in those cases where the enthusiastic cooperation of the employees is great enough for them to give their time freely.

Interdepartmental Competitions In Britain, a sure way of arousing employee interest is by means of competition with other departments. The first requirement is for some means of measuring performance. In a manufacturing department this is relatively easy, as the improvement can be measured in terms of the reduction in spoilt work, in the time spent on rectification, in the reduced number of customer complaints. In the drawing office, some companies have recorded the number of corrections which need to be made to detail drawings. The success of a design office, or of a development engineering department is far more difficult to reduce to finite terms. Especially is this true in the case of mechanical design, where there is so much art, and where ultimate success is not attained until customer satisfaction has been demonstrably achieved. Attempts have been made in clerical departments to count improvement in terms of the errors in typing, or in the number of documents requiring to be replaced.

In these latter cases, however, it is to be doubted if the effort to devise and apply a method of measurement can be justified by the improvements gained. Even Lord Kelvin would not insist that performance in these areas should be reduced to figures.

For this reason, it is not easy to compare performances between departments involved in different types of activity. Hence, it is more usual to confine competitions to departments where comparative measurements can be made.

During its trial run and again during Q & R Year, the Lucas organization maintained a running competition between production departments. Points were awarded each week to the department with the best performance. Results were displayed in a manner similar to that of a football league table, and an attractive incentive was provided which took the form of a continental holiday for each member of the winning group, with their families. Effective publicity ensured that the operatives received encouragement from their wives and children.

This scheme, and others of this company, received direct personal encouragement and inspiration from the chief executive, Sir Bertram Waring, who had long appreciated the value of Q & R and the importance of employee participation. It is known at first hand, however, that not all of his senior executives were equally enthusiastic at the outset. It was remarkable to see, however, how those who started out sceptically, feeling that the whole idea was a gimmick, were first surprised, then gratified, and finally converted by its success. It goes without saying that great care was taken in presenting the schemes to the employees. But it was a big help to know that they were what the chief wanted, and that they had his backing. Since his ideas were sound, too, and the result of a great deal of study over many years, they succeeded as well as they deserved to.

Other companies followed similar patterns, the difference lying mainly in the magnitude of the awards. This in turn depended upon the importance attributed by management to the degree of involvement of its employees.

Posters A little lower down the scale have been competitions for the best-designed posters to arouse interest in Q & R.

Great ingenuity has been shown in these, and they have certainly had a beneficial effect. Although they may have done more to identify which employees had poster-designing ability, they have certainly made the rest realize that the management was anxious to get them interested.

Suggestion Schemes These of course have been in existence for far longer than there have been Q & R promotions. Nevertheless, the Q & R activities have undoubtedly helped to revive suggestion schemes which were flagging.

Any motivational scheme requires to be refreshed from time to time, and focusing attention upon quality achievement has in some cases increased the participation in existing suggestion schemes by 50 per cent or more.

Zero Defects The 'ZD' scheme, which was initiated in the electronic based sector of the American aerospace industries a few years ago, has probably been the most widely publicized of all motivational methods.

It was started at a time when the then Secretary of Defense, McNamara, was urging that nation-wide efforts should be made to achieve maximum efficiency and economy in defence spending. In consequence, the scheme received the enthusiastic support of both military and civilian officers of the Department of Defense. Indeed, great moral pressure was applied to major contractors to the Government to adopt a ZD scheme.

Since useful conclusions can be drawn from an examination of this, probably the most widely applied of all promotional schemes, it will be worth while to examine it in detail.

252

'ZD' was conceived by Philip Crosby, when he was employed with the General Electric Co. The basic idea was, initially, to involve all employees in a total effort to reduce their errors as far as possible towards zero. The campaign was launched with all the panoply of which the United States is capable, with the Town Band and the Mayor participating. Possibly because the first attempt was made in a basically engineering industry, and in a region, New England, which had a long established manufacturing tradition, the scheme was not as successful as it might have been. It got off to a much better start, and became widely known when it was adopted by the Martin Orlando Co., situated in an orange-growing locality in Florida.

Here, there were significant differences. This company was involved in the manufacture of missiles for the US Army. Its products were primarily electronic. It was a light industry, established in a predominantly non-manufacturing area. The launching of the scheme coincided with a period when there was the need for increased business, and when Defense Secretary McNamara's economy drive was at its zenith.

The scheme was adopted with enthusiasm by the firm's senior management. Its promotion was carefully planned in detail. Every employee was required to pledge himself, over his signature, to devote all his efforts to the bettering of his individual performance, and in return he received an attractive badge bearing the letters ZD, which he wore at all times to show that he was not a non-conformist. In retrospect, it would appear that the management's task was made easier by the fact that the plant and the personnel were new, so that the introduction of a new plan did not encounter the difficulties of entrenched attitudes. And as some cynics have commented, those who did not wish to sign the pledge always had the option of going back to picking oranges.

The scheme was successful in reducing errors and in improving performance in all work areas, clerical, technical, and administrative as well as manufacturing. It was greeted with enthusiasm by the Defense Department in Washington, and considerable pressure was brought to bear upon contractors to adopt similar schemes. The large firms in the electronics and aerospace industries, many of which had expanded rapidly to meet the enormous procurement demands of the Department of Defense, were the first to respond. Officers of the three services adopted ZD with almost religious fervour, and visited participating companies to present awards to employees, a flag to the firm.[3]

Considerable sums of money were spent on promotion. This caused companies in older established industries, notably in those parts of the country which had long been industrialized, to express doubts about the universal applicability of the scheme. As their performance was already good, it was difficult to see that there would be a worthwhile return on the money which would be required to launch a scheme on such a scale as would impress the customer.

As experience with ZD schemes was accumulated, it became apparent that action which was beyond the scope of the operative was necessary to enable maximum performance to be achieved. Events were moving into the area of 'management-controllable defects'. Some firms adopted the slogan of Error Cause Removal, which has considerable merit as directing attention where management can do most good.

The emphasis of ZD has changed. Its primary objective is no longer solely to persuade the operative to produce better work. It is claimed to be predominantly a means of motivating management. This would appear to be a somewhat indirect approach, however. It involves putting pressure upon the worker to react by showing that he needs the help of management to achieve best results.

Impressed by the success stories which have crossed the Atlantic, some British firms have adopted ZD, sometimes without due consideration. Some have been successful, despite the disadvantage that the British pronunciation of Z is zed and not zee as in the United States.

Each management must make up its own mind on the course which might be best suited to the circumstances and the personality of its company. It is to be doubted, however, if, despite the successes which it can claim, ZD is better than the Q & R schemes which have been described. ZD has an overriding drawback, that it tends to be inward looking, whereas Q & R focuses attention upon the overall aim of satisfying the customer. This, it is thought, leads to the more direct approach. It is likely, too, to be more acceptable in companies where experienced employees would be sceptical about a non-attainable *zero*.

QC Circles Another scheme which has received a good deal of publicity is the QC Circle movement in Japan. Groups of shop floor operatives—in one case five girls about 20 years old—stay late without pay to work out solutions to problems which are causing spoilt work, or customer complaints. As a means of arousing employee interest, the scheme has its advantages. Shop floor workers are even trained in diagnostic method. One might question, however, the effect of the scheme upon the organization as a whole. In some of the cases quoted, the responsibility should have been placed clearly upon the designer, or upon the production engineer. It should be noted that these are deliberate attempts to eliminate known difficulties.

In some Japanese companies, the title of 'QC circle' is applied to a quite different type of group from that which signalized the start of the

254

movement. In these firms, the QC circle is no more than a team of responsible officials from technical departments, not significantly different from the committees of EMI (Electronics) Ltd described on page 241.

It is to be doubted if similar teams, which already exist in many British firms and which could with advantage be established in many more, would feel happier if they were expected to operate under a QC banner.

A Return to Earth

The pragmatic British manager will have become convinced of the value of the involvement of personnel in the aims of the enterprise. He may yet remain sceptical of the applicability to his own particular case, of the means just described.

He will welcome the advice of Patrick Fisher, of the Trades Union Congress, with whom we may well leave the last word:

> We shall be making a big mistake if we think that the desire for satisfaction in people's jobs is diminishing . . . define exactly what is wanted from your worker, what is meant by quality and reliability.[4]

References

1. Pearson, Sir Denning, welcoming a team of the Japanese Union of Scientists and Engineers, Derby, October, 1968.
2. 'Profiting by quality and reliability.' Report of National Conference 1966—with 41 Case Studies. National Council for Quality and Reliability.
3. Halpin, James F. *Zero Defects*. McGraw-Hill Book Company, 1966.
4. Fisher, Patrick. 'Quality and Reliability' Points for Trade Union Speakers. National Council for Quality and Reliability, 1967.

Appendix 1
The Quality Policy and Plan of Metalurgica de Santa Ana, S.A.

METALURGICA DE SANTA ANA, S.A.
QUALITY POLICY OF PRODUCTS

1.—*QUALITY POLICY OF M.S.A.*

The quality problem affects the whole Company: Quality of today is the guarantee of tomorrow.

To improve our position with regard to competition, and to increase the profits and reputation of our Company, it is necessary to manufacture products with sufficient quality to satisfy the requirements of the market with respect to the sales price.

The Board of Directors of M.S.A., has adopted the policy of putting on the market only those products having sufficient quality to deserve and achieve customers satisfaction, throughout the effective and reliable performance and during the designed life, of the functions required by the market.

2.—*DECLARATION OF PRINCIPLES*

To reach this objective, the quality levels of the products must be specified, and a policy must be drawn to regulate and limit the different functions of the Company directed to reach the said quality levels at the minimum possible cost.

The quality level will, at first, be determined by the specification and quality levels of our licensors, except those which are to be modified by the requirements of our market. For products of our own design the quality level will be that obtained by similar products made by world leading companies.

256

3.—*LINES OF ACTION AND RESPONSIBILITY*

3.1.—*Selection of products for the sales catalogue*

3.1.1.—The Commercial Management is responsible for specifically determining the characteristics of the product required by the market establishing its commercial features and minimum level of quality required for the optimum sales price.

3.1.2.—The Board is responsible for choosing only those products which will satisfy the market and which can be profitably manufactured by M.S.A.

3.1.3.—With regard to possible agreements for the manufacture of parts for other companies, the same rule will be applied in order to meet the customer's minimum requirements.

3.2.—*Technical specifications and documents*

The Engineering Executives are responsible for:

3.2.1.—Issuing the documents with full and clear specifications required to completely define the product in accordance with the quality levels set down in this policy.

3.2.2.—Not making reference to other documents in drawings or specifications other than those M.S.A. ones which are officially·in effect.

3.2.3.—Publishing all the complementary specifications required to secure the correct assembly of our products.

3.2.4.—Obtaining from our licensors the complete definition of the products to be manufactured and clarifying any doubt which could endanger the achievement of the quality levels set down in this policy.

3.2.5.—Bringing up-to-date the modifications drawn by our licensors, and supervising their introduction in accordance to adequate systems so that our products maintain the quality levels set down in this policy.

3.2.6.—Avoiding alterations of the original specifications of our licensors. If, for special circumstances of the market, sales or stock, this type of alterations is necessary, it should always be previously approved by our licensors. As soon as such circumstances will allow it, the specifications should be brought into line with the original ones as soon as possible.

3.2.7.—For products of our own design, not considering a

project as definitive until it has been checked that the results of the product, are of the specified quality.

3.2.8.—Not allowing drawings or documents, the quality level of same has not been proved, to be sent to other departments for approval without making clear that they are provisional.

3.3.—*Programming of new products*

Provisions for reaching the acceptable quality levels should be taken into consideration in the programming of new products.

The Executives involved in the programming and launching of a new product (or alternations of an existing product) have the responsibility of taking the necessary steps to obtain the specified quality levels in the time set by the programme.

3.4.—*Purveyance of materials and elements from sources outside M.S.A.*

The Purchase Management has the following responsibilities:

3.4.1.—To be fully familiar with the technical specifications which define the quality of materials and parts obtained from external sources.

3.4.2.—To only choose those suppliers capable of obtaining the specified quality levels and supply same with all the information required.

3.4.3.—To place no firm orders until the necessary samples have been accepted by the Quality Control Department.

3.4.4.—To stock materials and elements of the specified quality, allowing time for quality control before the required date of completion.

3.4.5.—To place orders taking into account possible rejects so that concessions need not be requested in order to meet the specified completion dates.

3.4.6.—To avoid the arrival to the factory of materials or parts which do not fulfil the technical specifications particularly in the Receiving Inspection Department, unless this deficiency could be known or ascertained by the Purchase Management.

3.4.7.—To refuse for storage materials or parts which have not been previously accepted by the corresponding Quality Control personnel, by written acceptance without control or by quality control procedures.

3.4.8.—To inform the supplier immediately of defects encountered and have them take corrective action.
3.4.9.—To eliminate suppliers whose products do not meet the specified levels of quality.
3.4.10.—To supervise transport and maintain the storehouses in such a condition that the materials and parts do not suffer damage.

3.5.—*Work programme*

The Programming Executives are responsible for:
3.5.1.—Foreseeing delivery dates, allowing time for the proper inspection and quality control of the finished parts, elements or products.
3.5.2.—Estimating the size of the lots, taking into account possible rejects so that concessions need not be requested in order to meet the completion programming dates.
3.5.3.—Refusing for storage, neither sending to following phases of production, those parts or elements that have not been accepted by Inspection in those phases where quality control is applied.
3.5.4.—Performing the necessary handling operations and keeping the warehouses in such a condition that the elements or assemblies do not suffer damage that would affect their quality.

3.6.—*Manufacture and assembly processes*

The Planning Executives of the different factories have the following responsibilities:
3.6.1.—To be fully familiar with the technical characteristics of the parts, assemblies and products which manufacturing and assembly processes they have to prepare, allowing suitable periods for quality control by production personnel.
3.6.2.—Translate the specifications into processes and production planning sheets for the work shops, designing, providing and allocating machines, tools, jigs, fixtures, space and proper time to obtain the specified quality levels.
3.6.3.—To design, provide and allocate the equipment, tools, gauges, means of testing, space and necessary time to measure the quality obtained by the manufacturer, as well as the corresponding operating instructions.

3.7.—*Manufacture and assembly*

The Production Executives of the different factories have the following responsibilities:

3.7.1.—To be fully familiar with the necessary processes and instructions for the manufacture, assembly and control, in order to obtain the specified quality levels.

3.7.2.—To manufacture parts and elements, and assembled units and products following the processes and instructions established, in order to obtain the specified quality levels.

3.7.3.—To apply the processes and established instructions to measure and determine the quality of the production obtained.

3.7.4.—Not to present to Inspections for their control, parts, elements, assemblies and products not having the quality levels specified in the technical documents.

3.7.5.—To maintain the required order and cleanness of the elements, assemblies, products, installations and workshops, in order not to lessen the specified quality levels.

3.7.6.—To assemble only and exclusively those parts which have been marked in each case as accepted by the Inspection Department.

3.8.—*Quality assurance*

The Quality Control Management has the following responsibilities:

3.8.1.—Establishing control processes, designing, providing and allocating the equipment, tools, gauges, means of testing, space and necessary time to assure an outgoing quality level acceptable to the market at the minimum control cost.

3.8.2.—Applying the proper control processes and inspection planning sheets so that the Company's products have the outgoing quality levels specified in the processes.

3.8.3.—Marking the parts and assemblies accepted which have functional importance. Marking all the products accepted.

3.8.4.—Checking the conditions of tools and gauges from all the factories and replacing those unserviceable, so that all factories have available the necessary elements in perfect condition at all times.

3.8.5.—Appraising and certifying the quality level reached by our Company by means of audits applied to materials, parts, assemblies of products accepted by Inspection and from the results of our products in the market.

3.9.—*Disposition of material, elements or products that are out of tolerance*

3.9.1.—The Purchase Management, with regard to materials, elements or products received from outside sources, has the following responsibilities:

3.9.1.1.—To have the material repaired if accepted with this condition and present it again to the Receiving Inspections.

3.9.1.2.—To return to the suppliers or destroy the material definitely rejected by the Receiving Inspections, during the manufacture or assembly process, or during the warranty period.

3.9.2.—The Production Executives have the following responsibilities:

3.9.2.1.—To repair the elements or products of own manufacture that have been accepted with this condition and present them again to the Final Inspections for their control.

3.9.2.2.—To destroy the elements or products of own manufacture which have been definitely rejected.

3.9.2.3.—To return to the Receiving Storehouses within the time fixed by the Purchase Management the materials and elements stocked from outside sources, that, having been accepted by Receiving Inspections, are rejected during the manufacturing process due to causes imputable to the suppliers.

3.9.3.—The Service Executives have the following responsibilities:

3.9.3.1.—Sending to the Quality Control Department the assemblies or elements received from our dealers, dismantled during the warranty period or in subsequent service, which may need investigation due to the importance or recency of the complaint.

3.9.3.2.—Destroying the assemblies or elements of our own manufacture claimed by our dealers and which do not require investigation.

3.9.3.3.—Returning to the Storehouses the assemblies or elements of outside manufacture claimed by our dealers, for their subsequent return to the suppliers by the Purchase Management.

3.9.4.—The Quality Control and Engineering Executives have the following responsibilities:

3.9.4.1.—Not making any concession that affects the security of the persons.

3.9.4.2.—Not making any concession that may lessen the reliability of our products.

3.9.4.3.—Requesting prior approval from our licensors of concessions which may be required due to their importance.

3.9.4.4.—Notifying the Service of those concessions required due to their importance.

3.10.—*Marketing*

The Commercial Executives have the following responsibilities:

3.10.1.—To avoid that the products received from the factories may suffer deterioration in quality during handling, storage and transport, and during storage by our dealers.

3.10.2.—To ensure that our dealers and customers know the advantages and limitations of our products as well as the necessary instructions for its handling and maintenance, and to avoid their use in markets or services for which they have not been designed and where they could be the cause of discredit for our Company.

3.10.3.—To avoid in commercial publicity exaggerations that could discredit our Company's reputation for quality.

3.10.4.—To know and check the quality levels of competitive products and the reaction of the market as regards the quality of our products. To pass on this information objectively to the persons concerned in the Company.

3.11.—*After-sales service*

The Technical Service and Spare Parts Executives have the following responsibilities:

3.11.1.—Guaranteeing that the authorized Service Workshops
have the necessary experience for properly servicing
our products, besides the necessary equipment, staff,
tools and instructions.

3.11.2.—Supervising that the product is properly serviced before
it is used.

3.11.3.—Refusing for storage in the Spare Parts Warehouse of
material which has not been accepted by Inspection.

3.11.4.—Avoiding the use of illegitimate spares by authorized
Service Workshops.

3.11.5.—Carrying out the handling and keeping the Spare Parts
Warehouses in such condition that the spares cannot
suffer quality damage and ensuring that the spares do
not suffer deterioration in the storehouses of the
authorized Service Workshops.

3.11.6.—Ensuring that the warranty claims' system is administered for the best service of both our Company and
customers.

3.11.7.—Turning over the information obtained from the
warranty claims in an objective and well documented
way to the functional groups concerned, so that it may
serve as basis for an efficient corrective action.

3.11.8.—Knowing the results of our products during their
service life and passing on this information objectively
to the functional groups concerned.

3.12.—*Corrective action for improving the quality level*

3.12.1.—The Quality Control Management is responsible for
checking, selecting and indicating the Departments or
functional groups which should correct the deviations
of the quality levels of our products during their
service life.

3.12.2.—All the functional groups are responsible for correcting
the deviations of the specified quality levels which may
occur in their respective working areas.

3.12.3.—The Quality Control Management is responsible for
pursuing the corresponding corrective action of each
functional group.

3.13.—*Quality cost*

The Finance Management is responsible for providing the
Quality Control Management with the necessary cost data to

direct the quality administration towards profitable and competitive results.

3.14.—*Practical working rules*

All the Company's Executives have the following responsibilities:

3.14.1.—Making this quality policy known to all persons who form part of their respective functional groups, specially those points which particularly refer to their specific functions.

3.14.2.—Motivating the persons who form part of their respective functional groups in order to reach a real quality ambiance in the Company where the quality level set by this policy be the direct and immediate result of the personal work of each individual, and not of the control and consequent corrective actions.

3.14.3.—Draft and publish the necessary working standards and instructions to put in practice this quality policy.

3.14.4.—Maintain a continuous programme which will permit reaching and maintaining, at the minimum possible cost, the objectives determined by our quality policy. These action programmes will have complementary programmes of necessary means approved by the M.S.A. Management.

Appendix 2
Techniques for the Manager

Quality and reliability have suffered at least as much as other management technologies through overemphasis upon techniques which have been little understood. Indeed, because most people have what Gossett called 'a popular dread of mathematical reasoning'[1] it is probable that Q & R techniques have suffered more than most.

There can be no gainsaying the value of the logic, and in many cases of the techniques, which have been evolved by applied statisticians. Unfortunately, these techniques have been taken up by mathematicians and others who have had either no wish or no opportunity to understand the circumstances and the needs of those people in industry who could benefit most. As a consequence, there tends to be a big gap between the purveyors of statistical techniques on the one hand, and the experienced industrial practitioners on the other, and both parties lose.

Many managers have been puzzled, and therefore suspicious, when some member of their staff has taken the trouble to pick up statistical techniques, in the belief that they will help in the solution of the company's problems. The individual concerned will most probably have become infected with the extreme enthusiasm which is typical of would-be industrial statisticians. As a result, he will tend to become a cultist, and his judgement will tend to be impaired.

Since some of the statistical techniques are the basis of methods of investigation and control which can be extremely valuable, it is important that management should have an understanding of their scope.

The dread of mathematics which was inculcated in most of us when we were at school can be diminished by realizing that the men who pioneered the methods were ordinary mortals. Gossett was a brewer, Shewhart a physicist, Deming graduated in agricultural engineering.

Only E. S. Pearson, who made a great effort shortly before the Second World War to bridge the gap, is a pure statistician, and his efforts were handicapped by the fact that the set type of his valuable work[2] was destroyed by bombing in 1940.

In this chapter, a brief account will be given of the scope of some of the more commonly used techniques. This should enable management to appreciate their value and to discount some of the more extravagant claims which are made by their enthusiastic advocates.

The Basis of Applied Statistics

By and large, the methods of control and investigation which are of most use to industry consist of no more than identifying the pattern of behaviour of a process, and comparing it with the pattern which it is known should usually apply. From this comparison, certain deductions can be made. If the patterns are similar, it can be assumed that the process is working normally. If they are not, the difference is due to some special or abnormal cause, which can then be sought out and rectified.

Such an oversimplification will probably be resented by statisticians. The remedy is open to them to do what is long overdue—to present in simple, understandable terms to their market what statistics are all about.

All products and processes vary under the accumulated influence of minor factors which cannot be eliminated or controlled economically. Such variations follow the same basic pattern. This pattern can be represented by a mathematical curve which is known as the normal distribution. This curve has certain well-defined characteristics, which can be put to good use.

It is a simple matter to determine the actual variability which is occurring in a particular process and to compare it with the normal, and so to see how the process is behaving.

The simplest applications of this normal distribution are to the determination of the capability of a process; to set up a process so that it will produce work within the desired tolerances; to keep the process 'in-control', i.e. working correctly. These are essential steps in the effort to get a job 'right first time'—to establish and control a process so that it does not produce unsatisfactory work.

Many courses are held in technical colleges throughout the country which teach the statistical techniques of quality control. Firms which operate continuous processes, for example those in the chemical industries, have long appreciated the value of statistical methods of control. Since it is desirable that their processes should operate within fairly close limits, the more sophisticated methods are applicable. Mechanical

266

engineering, however, is not such a precise industry, and less sophisticated methods would be more appropriate. Unfortunately, few of the available courses treat the subject at the level which is more suitable for the great majority of industrial problems. In consequence, large numbers of firms are deterred from profiting by simple techniques which could be picked up by anyone.

A few examples will illustrate these, and perhaps help some firms to reap benefit.

The Control Chart

This is a device which enables a visual record to be kept of the running performance of a job. It was conceived by Shewhart, and Fig. 3.1 reproduces his internal memorandum announcing it. Specialists have made it a more sophisticated method than is necessary in the great majority of cases.

When Deming first explained its working to a gathering of Japanese executives, he did so in simple terms, by means of the sketch shown in

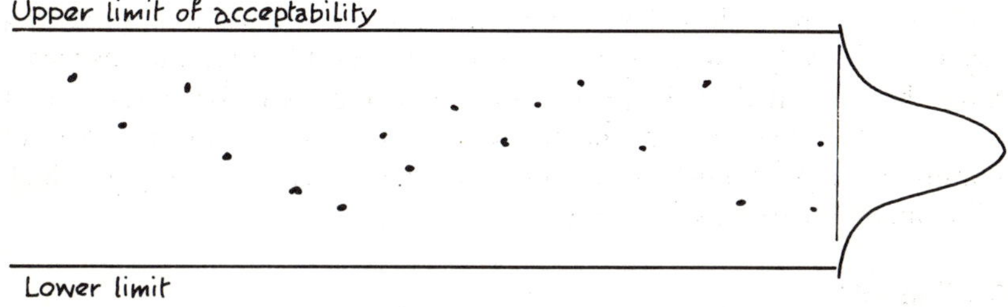

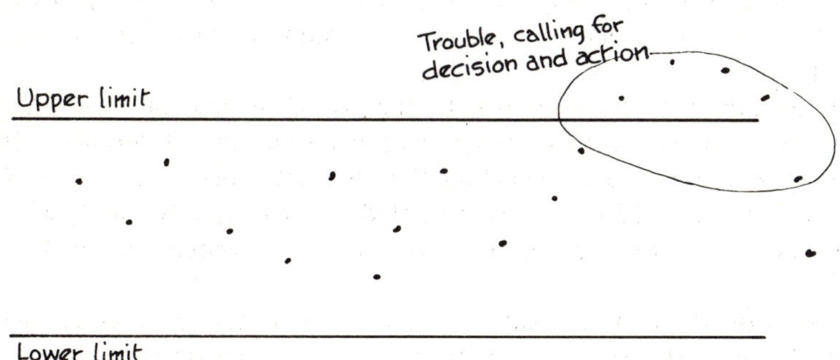

Fig. A2.1. Deming's Simple Control Chart.

Fig. A2.1.[3] This is all that is necessary for the great preponderance of industrial problems. All that is needed is that the production operative should be provided with a chart which indicates upper and lower control limits, which might for example be the upper and lower tolerances of a dimension being produced by the machining of a part. Or they might be the maximum and minimum weights of a filled packet of flour. Measurements are taken at specified intervals of time. A departure from the normal pattern is a signal that action should be taken to stop the process, or to discover and eliminate the cause of the deviation.

It is a pity that this was ever christened a *control* chart, because the adjective has had the unfortunate effect of causing the operative to suspect that it was he who was being controlled, and not the process. It has been suggested that a happier title would have been a 'How'm I doing' chart.

Preferably, the chart should be filled in by the operative himself. When this is possible, he becomes directly interested in his performance, which improves in consequence. When the rate of output does not allow the operative the time to measure and record the samples of product, the chart should be maintained by an itinerant or travelling inspector. It will now be seen why it is preferable that the inspector should be a member of the production department—he becomes 'one of us'. Also, the executive in charge, who is responsible for the costs of running his department, will be more willing to accept a method which will enable him to make savings.

Pre-Control

This is a simple technique which enables the capability of a process to be estimated rapidly. It provides, as well, a method of setting up a process so that it will be on target, and a means of maintaining control of the running process.

It can be explained by reference to the common case of a product which is being machined to dimensions, within specified tolerances. It depends upon known facts regarding the characteristics of the normal distribution curve. These are represented approximately, and with sufficient accuracy for the majority of industrial processes, in Fig. A2.2.

The assumption is made that the capability of the process is just adequate to meet the specified tolerances. This is pessimistic since if there were no margin of capability over the required tolerance the slightest shift from the central target, the slightest increase in variability, would result in parts being produced out of tolerance. The method, therefore, errs on the safe side.

268

A simple method of machine setting and control

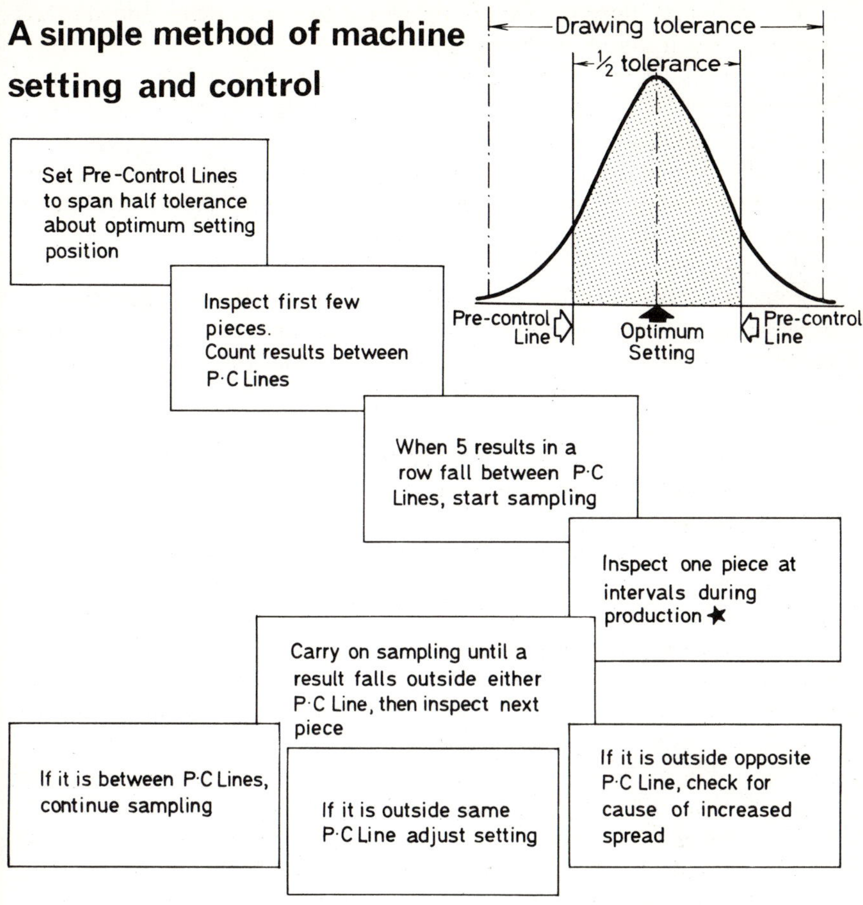

*Samples should be taken at intervals which give about 25 inspections for each machine correction.

Fig. A2.2. Pre-Control.

'Pre-control' lines are drawn representing half the total tolerance. It must be made clear to the operative that this is *not* depriving him of the full tolerance allowed to him. It is merely providing him with warning boundaries.

According to the known characteristics of the normal curve, provided that the operation is properly centred, and is behaving normally, 12/14 of all the products will lie within the half-tolerance or PC lines. One in 14 of them will be expected to fall in one or other of the outer areas.

The technique depends simply on assessing the chances that finished workpieces will fall in any particular area.

The process is described adequately in Fig. A2.2 and its use can be recommended as a means of demonstrating to production personnel a sound means of improving their performance.

Where greater accuracy is required, a simple method described by Jones,[4] which needs nothing more than plotting and simple arithmetic, can give useful results.

Pre-control was developed by a team of Rath and Strong Inc. A member of this team, Dorian Shainin, is one of the few men who is actively promoting a better understanding of simple 'strategies', which are in effect rule-of-thumb techniques derived from statistical principles.[5]

Acceptance Sampling

Early in the life of the quality assurance department of the Bell Telephone Laboratories, two of its members, H. F. Dodge and H. G. Romig, saw the value of sampling as a means of obtaining an estimate of the quality of a batch of product. This was an extension of the earlier work of Gossett, who had pioneered sampling as a means of estimating the quality of fields of barley intended for Guinness's stout. Unfortunately, means have become confused with ends, and sampling schemes have been refined far beyond the needs of most of industry. In addition, exaggerated claims have been made which have done little to inspire confidence, or to spread a proper understanding of the value of sampling.

For instance, it is often stated that 'statistical sampling is more effective than 100 per cent screening inspection'. This is manifestly untrue, and it is refuted by the protagonists themselves, who stipulate that when critical characteristics are involved which could affect the safety of a 'lethal' component, then there must be 100 per cent inspection.

Inspection by sampling depends upon the fact that a statistically-determined size of sample can indicate, with a desired degree of confidence (i.e. risk), what is the probable goodness of the batch as a whole, from which the sample has been taken.

Sampling can do no more than this. The technique can usefully be applied both as a final check that products which have been made under conditions of in-process control are as good as expected, and as a check that incoming goods are to the level (the 'acceptable quality level' or AQL, or, better the AOQL or 'average outgoing quality level') established by the supplier and accepted by the customer.

When sampling is used as a criterion for the acceptance of batches of goods, whether made in one's own plant and being checked by quality audit, or coming in from an outside supplier, it is essential that when a

270

sample indicates that a batch falls short of the desired quality, the whole batch should be rejected. If possible, the batch should be returned to the maker for correction by him. When this is not possible, because the parts are urgently needed in order to maintain a delivery programme for example, the batch should be 100 per cent checked by the customer, and the cost of this check should be charged to the supplier.

A difficulty which is often experienced is the determination of the AQL which will be 'acceptable' for each particular characteristic. It is the common practice to classify defects into categories such as:

CRITICAL DEFECTS Those which cannot be tolerated because they involve safety of life or because they will prevent the functioning of the product.

An AQL lying between 0·010 and 0·4 is usually selected, when sampling is used to monitor batches of parts which should already have had an effective 100 per cent inspection by the manufacturer.

MAJOR DEFECTS Those other than critical, which are likely to result in failure or materially to reduce the useability of the product.

AQL's between 1·0 and 2·5 are usual for this category.

MINOR DEFECTS Those unlikely to reduce the useability of the product.

AQLs in this category are usually between 6·5 and 10·0.

It must be pointed out that when the use of AQLs was advocated by the US Department of Defense, contractors were warned not to accept the AQL as a permissible level of defective quality. It is now being realized by Department of Defense officials that this warning has not been effective. There is a tendency not to quote AQLs but instead to measure the quality performance of a contractor. If this is acceptable, then acceptance sampling inspection is used to confirm that the contractor is maintaining his accustomed performance.

It will be readily apparent that the idea of accepting batches on an indication that they may contain a percentage of unsatisfactory parts is repugnant to most inspectors, engineers, and managers. What they must try to realize is that the sampling inspection is, or should be, no more than a monitoring check to indicate that the batch is to the customary standard.

Another difficulty, which is experienced especially by inspectors in mechanical engineering, who have been trained to reject all unsatisfactory products, arises with sampling schemes which indicate that a certain AQL will be achieved if no more than say 5 defectives are found in a batch.

The following extract from the export inspection scheme of the Japanese Camera Inspection Institute will illustrate this point:

The inspection is conducted by random sampling. The following table shows how sampling is carried out.

MECHANICAL OPERATING CHARACTERISTICS

N	n	p	r
Less than 110	20	1	2
111– 180	30	2	3
181– 500	40	4	5
501– 800	50	5	6
801–1300	60	6	7
1301–3200	70	8	9
3201–5000	85	10	11

N: Number of pieces forming a lot.
n: Number of pieces to be sampled.
p: Maximum number of defectives which allows the lot to be passed.
r: Number of defectives which causes the lot to be rejected.

Each camera in the sample, say of 40 cameras from a batch of 300, is subjected to tests of appearance and optical and mechanical performance. If more than 5 are found defective, the whole batch of 300 cameras is returned to the manufacturer. If there are 4 or fewer defectives, all the cameras receive the golden 'Passed J.C.I.I.' sticker, the 4 or so defective ones being made good.

This sample inspection represents an AQL of about 5 per cent. In other words, 5 per cent of the cameras in the batch can be expected to be defective in some respect.

The objection to acceptance figures of this order (i.e. accept on 4, reject on 5) is overcome by some companies, who choose the sample size so as to lead to acceptance/rejection figures of 0,1. This has the great advantage that an experienced inspector can all the more easily be trained to use the scheme with confidence.

The Ford Motor Co. Ltd have such a scheme, which is shown on page 273.

It is necessary to repeat the warning that sampling inspection needs to be fully understood if it is to be used as more than a means of monitoring and disciplining suppliers. Only recently have official bodies begun to issue specific warnings. The two which follow give sound advice:

SAMPLING PLANS Sampling plans may be used when inspection or tests are destructive, or data, inherent characteristics, or the noncritical application of an article or material indicate that a reduction in inspection or testing can be achieved without jeopardising achievement of quality, reliability, or design intent.[6]

Statistical techniques shall not be used to eliminate final inspection, but may be used to reduce final inspection providing the techniques and their application are satisfactory to the Ontario Hydro Representative.[7]

SAMPLING TABLE FOR LOT INSPECTION *Ford Motor Company Ltd.*

LOT OR SHIPMENT SIZE	MINIMUM SAMPLE SIZES				DEFECTIVE PIECES ALLOWED
	TABLE A	TABLE B	TABLE C	TABLE D	
1 to 149	15*	30*	75*	Inspect 100%	0
150 to 499	30	50	115	150	0
500 to 7,999	50	150	300	300	0
8,000 and over	115	300	500	500	0

MINIMUM REQUIREMENTS

Major Characteristics:
—Reduced Inspection.....................Table A
—Normal Inspection......................Table B
—Tightened Inspection...................Table C
Critical Characteristics.................Table D

*WHEN LOT SIZE IS LESS THAN MINIMUM SAMPLE SIZE, INSPECT 100%

QUALITY CONTROL
MANUFACTURING STAFF
JUNE 1964

Techniques for Process and Product Improvement

Excessive preoccupation with the application of statistics to quality control has diverted attention from other statistical techniques which have a far greater potential value to the design and development engineer, to the production engineer, and to the service engineer, and hence to the enterprise as a whole.

The development engineering department may be confronted with the task of discovering what are the factors which prevent the satisfactory functioning of a new design of product. A certain bearing may fail consistently, no matter what changes are made to the fit on the shaft and in the housing, to the cage clearances, to the design of the cage itself, to the running clearances. Or a seal may be leaking on an unacceptably high proportion of pump casings. Changing the section of the seal rubber, the size of the ring groove, the stiffness of the cover, effect no improvement. The changes and the necessary tests are costly and time consuming.

A more logical approach than the hit and miss methods which are usually practised depends upon the more efficient marshalling of data. This method has been evolved from the work of Fisher, who first described the use of designed experiments, analysis of variance and regression analysis as applied to biological research.[8]

Today's exponent of these methods, in terms which are understandable to engineers, is Shainin. He tackles a problem such as that of the pump seal by approaching it logically. Measuring the leakage from each of a consecutive run of pumps, he applies what he calls variation research. The measured leakages are arranged so as to show:

(a) The variation of leakage within the units, i.e. is there a pattern suggesting that the leakage occurs at a particular spot around the seal cover?
(b) The variation from unit to unit, i.e. is assembly a factor?
(c) The variation within batches made at different times, i.e. is there a factor as between the beginning and end of a shift?

By putting the measured leakages in the order which will reveal the answers to the questions posed above, it is possible to eliminate most of the causes which have been postulated, and, quite quickly, to arrive at the most likely cause of trouble. A simple test will then confirm that the conclusion is the right one. In the example quoted, it was found to be porosity in some of the cast covers.

Comparing this with the typical 'engineering' approach shows that it is much more efficient. Information which can be obtained at little cost, by measuring the characteristics of a run of consecutive units of product, can lead to the answer which solves the problem. The hit and miss method could require a number of modifications and tests. This method would take longer, and cost more.

Variation research uses, in effect, the technique which is known as the Latin square.[9] More complex problems, where a large number of factors may be involved, can be solved by a method known as 'factorial design'. Again, it is no more than a means of so arranging and numbering the

274

data so that it, the evidence, is made to reveal the factor, or the combination of factors, which have the greatest effect.[10]

When it is desired to confirm that a new design has been successful in producing a better performance than that of the existing design, we may have recourse to the hypothetical distribution evolved by Waloddi Weibull. This was first put forward by Weibull in 1954, but it was not until many years had elapsed that the value of Weibull's work was seen. As is usual, the 'discoverers' of this work endowed it with powers beyond its capabilities, and it took quite a time to establish the scope and limitations of this valuable method. By plotting lives at failure on a specially designed chart[11] it is possible to forecast, for a minimum number of life tests, whether or not the new product is likely to be an improvement, as far as reliability is concerned, over the existing design. As with so many statistical techniques, this Weibull method produces the best result when it is combined with engineering judgement.[12]

At the time of writing, little has been published describing the practical applications of this method. Shainin, however, has made, with the collaboration of the Pratt & Whitney Co., a number of films which present this and the other techniques which have been mentioned.[13]

Another valuable technique, which was evolved years ago by Box and Wilson, enables development engineers to optimize performance. The efficiency of a compressor, for example, or of an engine or a motor or a chemical process, might depend upon a number of factors, acting in combination. Evolutionary Operation, as this method is known, makes it possible to determine the precise combination of the relevant factors which produces the highest efficiency. This is done by recording the performance achieved and the prevailing levels of the controlling factors. By making small changes in these factors, which are within the variations which can be tolerated during normal operation of the process, it is possible to obtain an indication of the direction in which overall improvement lies. By moving progressively in this direction, the optimum combination can be arrived at, often without carrying out special tests, but rather by using current information and so ordering it as to make it divulge the maximum of useful information.[14] This method has been described in simple terms in Shainin's first film.[13a]

As has been indicated, there is a dearth of training in the application of statistical techniques to engineering problems. In companies where there is a degree of innovation—of design or of method of manufacture—management should certainly consider carefully the potential value of having at least one engineer trained in statistical method, and thus begin to reap the benefits of a better planned approach, which were pointed out to industry, by British pioneers, nearly half a century ago.

References

1. McMullen, L. and Pearson, E. S. 'William Sealy Gossett, 1876–1937.' *Biometrika*, January, 1939.
2. Pearson, E. S. 'BS 600, 1935.' British Standards Institution.
3. Deming, W. Edwards 'On the teaching of statistical principles and techniques to people in industry.' *Bulletin*, International Statistical Institute, Tome XXXIV. Rome, 1954.
4. Jones, J. S. 'A graphical method of providing a visual presentation of the capability of a manufacturing process in relation to the specification requirements.' International Conference on Quality Control, Tokyo, October, 1969.
5. Putnam, A. O. 'Pre-control.' Section 19 of Juran's *Quality Control Handbook*. McGraw-Hill Book Company, 2nd Ed., 1962.
6. 'Quality Program Provisions for Aeronautical and Space System Contractors.' Publication NHB.5300 4(IB) April 1969. National Aeronautics and Space Authority. U.S. Government Printing Office, Washington D.C. 20402, U.S.A.
7. 'Information and Guidance for the Evaluation of Company Quality Programs.' Handbook Q.A.-20-69, Ontario Hydro, Canada.
8. Fisher, Sir Ronald A. *The Design of Experiments*. Oliver & Boyd, London, 1935.
9. Shainin, Dorian. 'The designed experiment.' *Harvard School of Business Review*, July–August 1957.
10. Shainin, Dorian:
    (*a*) 'Problem-solving math. made easy.' *Factory*, June 1959.
    (*b*) 'Strategies for solving manufacturing problems.' *Automation*, February 1966.
    (*c*) 'How to Achieve More Productive Experimentation.' Report published by Rath & Strong Inc., Boston, Mass. 02110, U.S.A.
11. Nelson, Lloyd S. 'Weibull probability paper.' *Evaluation Engineering*, May/June, 1967.
12. Bompas-Smith, J. H. 'The Determination of Distributions that Describe the Failures of Mechanical Equipment,' loc. cit.
13. These films are obtainable on loan from the Central Film Library, London:
    (*a*) *Increasing the Efficiency of Development Testing*.
    (*b*) *Some Reliability Considerations in a Development Program*.
    (*c*) *The Weibull Distribution*.
    Made by Dorian Shainin in collaboration with the Pratt & Whitney Company, U.S.A.
14. Box, G. E. D. and Wilson, K. B. 'Experimental attainment of optimum conditions.' *J. Royal Soc.*, Vol. 13, 1951.

Index

282

286

PRINTED AND BOUND IN ENGLAND BY
HAZELL WATSON AND VINEY LTD
AYLESBURY, BUCKS